PROBLEMS IN MODERN PHYSICS

PROBLEMS IN MODERN PHYSICS

WILLY SMITH

Department of Physics
Lycoming College
Williamsport, Pennsylvania

GORDON AND BREACH, SCIENCE PUBLISHERS

NEW YORK • LONDON • PARIS

Library of Congress Catalog Card Number: 74-11587

Editorial office for Great Britain:
Gordon and Breach, Science Publishers Ltd.
12 Bloomsbury Way
London W.C.1

Editorial office for France:
Gordon & Breach
7-9 rue Emile Dubois
Paris 14e

Printed in the United States of America

0287

P R E F A C E

This is a collection of problems in Modern Physics for undergraduate students who have already taken a sequence of two semesters in General Physics. The statement of each problem is followed by a detailed solution, the basic idea being to teach the student, through example, how to organize his thoughts and construct a logical path from the given information to the desired result. Freed from the preconceived idea that looking up the solution is "cheating," he will be able to focus his attention on the procedure and eventually build a satisfactory general approach to problem solving. A very few problems without detailed solutions, but similar to others already discussed, have also been included and their numerical answers provided.

The sources for the problems are numerous and diversified. The collection has been built over the years as a by-product of the teaching activities of the author. The ideas evolved from the particular needs of the course, from the necessity for emphazising some points that had seemed to be pitfalls for countless students, and from discussions with colleagues an d associates. Some of the problems are "classics" and have been handed down from textbook to textbook; some others are more or less standard, at about the level of the examinations presently given in many colleges and universities. The author does not claim the ownership of any but a few of the problems, but the solutions are his. The degree of sophistication of the problems, as well as the scope of the mathematical apparatus required for their solution, is also quite varied.

The subjects included are those usually encountered in a course in Modern Physics, although not all of them have been dealt with in equal depth, and perhaps some topics have been altogether neglected. This, however, reflects only the preferences of the author.

Although the trend in the literature seems to favor the CGS system of units regardless of the international efforts in support of the MKS system, no emphasis has been placed in this work on one particular system of units, since it will be to the advantage of the reader to be equally familiar with both systems. Atomic masses based on carbon 12, as well as those based on oxygen 16 have been used, since references for the latter are more abundant than for the former and therefore perhaps more readily available for the student at the undergraduate level.

The majority of the problems have been class-tested. However, it would be presumptuous to assume that there are no errors, typographic or otherwise. It is the author's experience that no matter how many times one looks at things, much to one's chagrin a mistake or a misprint always manages to slide through. Thus, the author will be appreciative if any error is reported to him. He will also welcome any criticism that could result in an improvement in this work.

The author wishes to acknowledge the effort and cooperation of Mr. John J. Yahner of the Williamsport Area Community College in the preparation of the many drawings included here.

WILLY SMITH

Williamsport, Pa.

To *German E. Villar*

who showed me the way.

CONTENTS

I

SPECIAL RELATIVITY

1

Expand the relativistic expression for the kinetic energy in powers of β and show that when $\beta \ll 1$ **the classical expression,** $K = m_o v^2/2$, is obtained. Determine the values of γ and the corresponding values of β for which the true kinetic energy is equal to: (a) 1.01 times the non-relativistic value; (b) 1.1 times, and (c) 5 times.

****

The relativistic expression for the kinetic energy is:

$$K = m_o c^2(\gamma - 1) = m_o c^2\{(1-\beta^2)^{-1/2} - 1\} \qquad (1)$$

But the binomial expansion of $(1-\beta^2)^{-1/2}$ is:

$$(1-\beta^2)^{-1/2} = 1 + \frac{1}{2}\beta^2 + \frac{-\frac{1}{2}(-\frac{1}{2}-1)}{2!}\beta^4 + \cdots\cdots$$

$$= 1 + \frac{1}{2}\beta^2 + \frac{3}{8}\beta^4 + \cdots\cdots$$

and (1) becomes:

$$K = m_o c^2(1 + \frac{1}{2}\beta^2 + \frac{3}{8}\beta^4 + \cdots - 1) = m_o c^2(\frac{1}{2}\beta^2 + \frac{3}{8}\beta^4 + \cdots\cdots)$$

In the classical limit, $\beta \ll 1$ and higher powers of β can be neglected:

$$K = m_o c^2 \frac{1}{2}\frac{v^2}{c^2}$$

or:

$$\boxed{K = \frac{1}{2} m_o v^2}$$

in agreement with the correspondence principle. QED.

Now, the ratio of the relativistic kinetic energy to the classical value is written:

$$\alpha = \frac{m_o c^2(\gamma-1)}{\frac{1}{2} m_o v^2} = \frac{m_o c^2(\gamma-1)}{\frac{1}{2} m_o \beta^2 c^2} = \frac{2(\gamma-1)}{\beta^2}$$

But: $\beta^2 = \gamma^2 - 1/\gamma^2$, and this becomes:

$$\alpha = \frac{2(\gamma-1)\gamma^2}{\gamma^2-1} = \frac{2\gamma^2}{\gamma+1}$$

or: $$2\gamma^2 - \alpha\gamma - \alpha = 0$$

Solving for γ :

$$\gamma = \frac{\alpha + \sqrt{\alpha^2 + 8\alpha}}{4} \qquad (2)$$

where only the + sign is used since $\gamma \geqslant 1$. Values of γ corresponding to the different values of α are calculated using this formula. Then, the values of β are obtained from the relation:

$$\beta = \frac{\sqrt{\gamma^2-1}}{\gamma} \qquad (3)$$

Numerical results are tabulated below:

α	γ	β
1.01	1.0065	0.115
1.10	1.068	0.347
5.00	3.2655	0.952

If tables of hyperbolic functions are available, the corresponding values of γ and β can be easily obtained by noticing that if one makes:

$$\beta = \tanh x, \quad \text{then:} \quad \gamma = \cosh x$$

See, for example, HANDBOOK OF MATHEMATICAL FUNCTIONS, NBS, AMS 55, 1964, pp.213-18.

*****

2

(a) How many times its rest mass is the actual mass of an electron which has been accelerated through a potential difference of 100 kilovolts ?
(b) What is its velocity in terms of c ?
(c) What would be the percentage error if the velocity is calculated using the classical formula, $K = m_o v^2/2$?

****

(a) Expressed in Mev, the rest energy of the electron is 0.511 Mev. Thus:

$$E = K + E_o = 0.100 + 0.511 = 0.611 \text{ Mev}$$

and since:

$$\frac{m}{m_o} = \frac{E}{E_o} = \gamma$$

one obtains:

$$\gamma = \frac{0.611}{0.511} = 1.196$$

or:

$$\boxed{m = 1.196\ m_o}$$

Answer (a)

(b) Now:

$$\gamma = \frac{1}{\sqrt{1-\beta^2}} \quad \therefore \quad \beta = \frac{\sqrt{\gamma^2 - 1}}{\gamma} = 0.549$$

or:

$$\boxed{v = 0.549\ c}$$

Answer (b)

(c) Using the classical expression for the kinetic energy, $K = m_o v^2/2$, one gets:

$$v^2 = \frac{2K}{m_o}$$

Dividing by c^2:

$$\beta^2 = \frac{v^2}{c^2} = \frac{2K}{m_o c^2} = \frac{2 \times 0.100}{0.511} = 0.391$$

or: $\beta = 0.625$

and: $v_{class.} = 0.625\ c$

The percentage error is therefore:

$$\varepsilon = \frac{100\ \Delta v}{v} = \frac{100(0.625c - 0.549c)}{0.549c} = \frac{100 \times 0.076}{0.549}$$

or:

$$\boxed{\varepsilon = 13.85\%}$$

Answer (c)

****

3 A photon has the same momentum as a 1 Mev electron. (a) What is the energy of the photon ? (b) What would be the frequency of this radiation ?

***

The total energy of the electron is: $E^2 = p^2c^2 + E_o^2$

but also: $E = K + E_o$

Combining these two expressions: $p^2c^2 = K^2 + 2KE_o$

where K, the kinetic energy, is in this case 1 Mev, while $E_o = 0.511$ Mev.

Thus:

$$p^2c^2 = 1^2 + 2 \times 1 \times 0.511 = 1 + 1.022 = 2.022 \text{ Mev}^2$$

or: $$pc = 1.42 \text{ Mev}$$

Since the photon has no rest mass:

$$\boxed{E_\gamma = pc = 1.42 \text{ Mev}}$$ Answer (a)

Now: $E_\gamma = h\nu \qquad \nu = \frac{E_\gamma}{h} = \frac{pc}{h}$

Hence: $$\nu = \frac{1.42 \text{ (Mev)} \times 1.6 \times 10^{-6} \text{(erg/Mev)}}{6.625 \times 10^{-27} \text{(erg -sec)}}$$

or: $$\boxed{\nu = 3.42 \times 10^{20} \text{ cycles/sec}}$$ Answer (b)

******

4

In a classical collision of a particle with another particle of the same mass at rest, the angle between the two trajectories after the collision is always $\pi/2$. Show that for a relativistic collision:

$$\tan\theta \tan\phi = \frac{2}{\gamma + 1}$$

where θ, ϕ are the angles of the out-going particles with respect to the direction of the incident particle, and γ corresponds to the incident particle before the collision. Show that $\theta + \phi \leqslant \pi/2$, where the equal sign is valid in the classical limit.

****

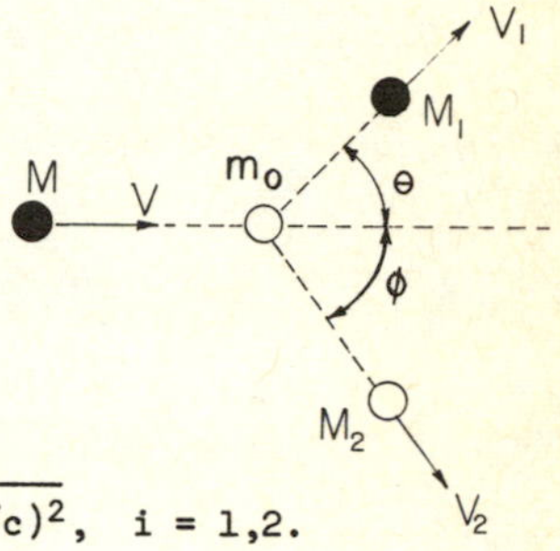

Using the notation of the figure, conservation of energy is written:

$$E + m_o c^2 = E_1 + E_2$$

or: $$m_o\gamma c^2 + m_o c^2 = m_o\gamma_1 c^2 + m_o\gamma_2 c^2$$

$$\gamma + 1 = \gamma_1 + \gamma_2 \qquad (1)$$

where:

$$\gamma = 1/\sqrt{1 - (v/c)^2}\,, \qquad \gamma_i = 1/\sqrt{1 - (v_i/c)^2}, \quad i = 1,2.$$

Conservation of momentum is expressed:

$$m\underline{v} = m_1\underline{v}_1 + m_2\underline{v}_2$$

or: $$m_o\gamma\underline{\beta}c = m_o\gamma_1\underline{\beta}_1 c + m_o\gamma_2\underline{\beta}_2 c$$

$$\gamma\,\underline{\beta} = \gamma_1\underline{\beta}_1 + \gamma_2\underline{\beta}_2$$

which yields two scalar equations:

$$\gamma\beta = \gamma_1\beta_1 \cos\theta + \gamma_2\beta_2 \cos\phi \qquad (2)$$

$$0 = \gamma_1\beta_1 \sin\theta - \gamma_2\beta_2 \sin\phi \qquad (3)$$

The angle ϕ is eliminated by squaring and adding the two momentum equations:

$$\gamma^2\beta^2 + \gamma_1^2\beta_1^2 - 2\gamma_1\beta_1\gamma\beta \cos\theta = \gamma_2^2\beta_2^2 \qquad (4)$$

Since: $\gamma_i\beta_i = \sqrt{\gamma_i^2 - 1}$, above equation becomes:

$$\gamma^2 - 1 + \gamma_1^2 - 1 - 2\sqrt{\gamma_1^2 - 1}\,\sqrt{\gamma^2 - 1}\cos\theta = \gamma_2^2 - 1$$

The parameter γ_2 is now eliminated between this equation and the energy equation. From (1):

$$\gamma_2 = \gamma + 1 - \gamma_1$$

or: $$\gamma_2^2 = \gamma^2 + 1 + \gamma_1^2 - 2\gamma\gamma_1 + 2\gamma - 2\gamma_1$$

Thus: $$\gamma^2 - 1 + \gamma_1^2 - 1 - 2\sqrt{(\gamma_1^2 - 1)(\gamma^2 - 1)}\cos\theta = \gamma^2 - 2\gamma_1(\gamma+1) + 2\gamma + \gamma_1^2$$

or: $$2 + 2\sqrt{(\gamma_1^2-1)(\gamma^2-1)}\cos\theta = 2\gamma_1(\gamma+1) - 2\gamma$$

which simplifies to: $$(\gamma_1+1)(\gamma-1)\cos^2\theta = (\gamma_1-1)(\gamma+1)$$

and solving for γ_1:

$$\gamma_1 = \frac{(\gamma+1)+(\gamma-1)\cos^2\theta}{(\gamma+1) - (\gamma-1)\cos^2\theta}$$

Let: $$\frac{\gamma-1}{\gamma+1} = a^2$$

Thus: $$\gamma_1 = \frac{1 + a^2\cos^2\theta}{1 - a^2\cos^2\theta} \qquad (5)$$

Returning now to the momentum equations (2) and (3) and dividing them:

$$\tan\phi = \frac{\gamma_1\beta_1 \sin\theta}{\gamma\beta - \gamma_1\beta_1\cos\theta} \qquad (6)$$

But:

$$\gamma_1\beta_1 = \sqrt{\gamma_1^2 - 1} = \sqrt{\left(\frac{1 + a^2\cos^2\theta}{1 - a^2\cos^2\theta}\right)^2 - 1}$$

$$= \frac{\sqrt{(1 + a^2\cos^2\theta)^2 - (1 - a^2\cos^2\theta)^2}}{1 - a^2\cos^2\theta} = \frac{2\,a\cos\theta}{1 - a^2\cos^2\theta}$$

and:

$$\gamma\beta = \sqrt{\gamma^2 - 1} = \sqrt{(\gamma+1)(\gamma-1)} = \sqrt{a^2(\gamma+1)^2} = a(\gamma+1)$$

By substitution into (6), then:

$$\tan\phi = \frac{\dfrac{2a\sin\theta\cos\theta}{1 - a^2\cos^2\theta}}{a(\gamma+1) - \dfrac{2a\cos^2\theta}{1 - a^2\cos^2\theta}} = \frac{2\sin\theta\cos\theta}{(\gamma+1) - (\gamma+1)a^2\cos^2\theta - 2\cos^2\theta}$$

$$= \frac{2\sin\theta\cos\theta}{\gamma(1 - \cos^2\theta) + (1 - \cos^2\theta)} = \frac{2}{\gamma+1}\;\frac{\sin\theta\cos\theta}{1 - \cos^2\theta} = \frac{2}{\gamma+1}\;\frac{1}{\tan\theta}$$

Finally: $$\boxed{\tan\theta\tan\phi = \frac{2}{\gamma + 1}} \qquad \text{Q. E. D.}$$

<u>Discussion</u>: In the classical limit, $\gamma \to 1$, and $\tan\theta\tan\phi \to \frac{2}{2} = 1$.

But since: $$\tan(\theta+\phi) = \frac{\tan\theta + \tan\phi}{1 - \tan\theta\tan\phi}$$

then: $\tan(\theta+\phi) \to \infty$, i.e., $\theta + \phi = \pi/2$

Hence, for $\gamma > 1$: $$\boxed{\theta + \phi \leqslant \pi/2} \qquad \text{Q.E.D.}$$

*****

5

A π^+ meson at rest decays into a μ^+ meson and a neutrino in 2.5×10^{-8} seconds. Assume now that this π^+ meson has kinetic energy equal to its rest energy.

(a) What is the velocity of the meson ?

(b) What distance would the meson travel before decaying, as seen by an observer at rest ?

****

(a) Let: E_o = rest energy, K = kinetic energy, E = total energy.

Then: $E = K + E_o = 2E_o$

But also: $E = \gamma E_o = \dfrac{E_o}{\sqrt{1 - \beta^2}}$

$$\therefore \beta^2 = 1 - \left(\frac{E_o}{E}\right)^2 = 1 - \frac{1}{4} = \frac{3}{4}$$

or: $$\boxed{v = \frac{\sqrt{3}}{2} c = 0.866\ c = 2.598 \times 10^8\ \frac{m}{sec}}$$

(b) The decay time of the π^+ in his own frame of reference is:

$$T_o = 2.5 \times 10^{-8}\ sec$$

For an observer at rest, the meson is moving with speed $\frac{\sqrt{3}}{2}c$, hence: $\gamma = 2$,

and: $$T_1 = \gamma T_o = 2 \times 2.5 \times 10^{-8}\ sec = 5 \times 10^{-8}\ sec$$

The distance traveled before decaying is then:

$$d = vT_1 = \frac{\sqrt{3}}{2} c\ 2T_o = \sqrt{3}\ T_o c = \sqrt{3} \times 2.5 \times 10^{-8} \times 3 \times 10^8$$

or: $$\boxed{d = 13\ meter}$$

****

6 The trips of two space ships to nearby stars take place according to the following schedule:

	Space ship No.1	Space ship No.2
Acceleration period (to o.8c)	1 year	1 year
Outward trip at o.8 c	1 "	6 "
Deceleration, exploring and new acceleration period (to 0.8c)	1 "	1 "
Return trip at 0.8c	1 "	6 "
Final deceleration	1 "	1 "

All above times were recorded by the observers on Earth. Determine the difference in duration of the trips as shown by clocks carried in the space ships. Consider only actual trip duration, disregarding the acceleration and deceleration periods.

****

First, one notices that for $\beta = 0.8$ one has:

$$\gamma = \frac{1}{\sqrt{1 - \beta^2}} = \frac{1}{\sqrt{1 - 0.64}} = \frac{5}{3}$$

Now, the round trip for space ship No.1 took, as measured by observers on Earth:

$$T = \gamma T_o = 2 \text{ years}$$

Thus:

$$T_o = \frac{T}{\gamma} = \frac{3}{5} \times 2 = 1.2 \text{ year}$$

as measured by the clocks on board. Similarly, for the second space ship:

$$T = \gamma T_o = 12 \text{ years}$$

and:

$$T_o = \frac{T}{\gamma} = \frac{3}{5} \times 12 = 7.2 \text{ year}$$

The difference in duration is then:

$$\Delta T = 7.2 - 1.2 = 6 \text{ years}$$

****

7 At the time a space ship moving with speed $v = 0.6c$ passes the U.S. space station located near by the orbit of Mars, a radio signal is sent from the station to Earth. This signal is received on Earth 1250 seconds later.

(a) What is the duration of the trip from the space station to Earth according to the observers on Earth ?

(b) What is the duration according to the crew of the space ship ?

****

The distance from the space station to Earth is determined by the time required for the radio signal to reach Earth:

$$d = ct = 1250c$$

As seen by the observers on Earth, at $t_1 = 0$, the space ship is at a distance d and approaching with speed 0.6c. This, it will arrive after a time:

$$t_1 = \frac{\text{distance}}{\text{speed}} = \frac{d}{\beta c} \qquad (1)$$

or, numerically: $$t_1 = \frac{1250\ c}{0.6\ c} = \frac{1250}{0.6} = \boxed{2083\ \text{sec}}$$ **Ans.(a)**

Now, the clocks in the space ship will record a proper time t_2 related to t_1 as follows:

$$t_1 = \gamma t_2$$

or: $$t_2 = \frac{t_1}{\gamma} = \frac{d}{\gamma\ \beta c} \qquad (2)$$

Numerically, since:

$$\gamma = \frac{1}{\sqrt{1-\beta^2}} = \frac{1}{\sqrt{1 - 0.6^2}} = 1.25$$

one has:

$$t_2 = \frac{1250\ c}{1.25 \times 0.6c} = \frac{1000}{0.6} = \boxed{1666\ \text{sec}}$$ **Ans.(b)**

The same results should be obtained, of course, by using the Lorentz transformation. Considering an inertial frame S_1 fixed to the space station, and a second frame S_2 moving with the space ship, one has:

$$x_2 = \gamma(x_1 - vt_1) \qquad (3)$$

$$t_2 = \gamma(t_1 - \frac{v}{c^2}x_1) \qquad (4)$$

$$\dot{x}_2 = \frac{\dot{x}_1 - v}{1 - \frac{v}{c^2}\dot{x}_1} \qquad (5)$$

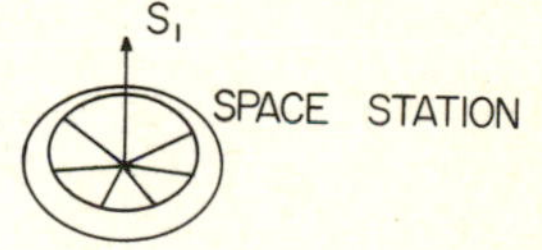

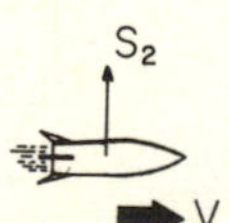

At $t_1 = t_2 = 0$, when the space ship is passing by the station, the origins coincide.
Since the distance from the space station to Earth is constant, Earth is at rest in frame S_1, and its space-time coordinates are:

$$x_1 = d\ , \qquad t_1 = 0$$

The space-time coordinates of Earth in frame S_2, using (3) and (4), are then:

$$x_{2E} = \gamma d, \qquad t_2 = -\frac{\gamma v d}{c^2}$$

and Earth is approaching the space ship (at $x_2 = 0$) with velocity $\dot{x}_{2E}$ obtained from (5) for $\dot{x}_{1E} = 0$:

$$\dot{x}_{2E} = -\beta c$$

The negative sign indicates that in S_2 Earth is moving from right to left. The time required to reach the origin is then:

$$\Delta t = \frac{x_{2E}}{\dot{x}_{2E}} = \frac{\gamma d}{\beta c}$$

and the clocks in the ship will read then:

$$t_2 = -\frac{\gamma v d}{c^2} + \frac{\gamma d}{\beta c} = \frac{\gamma d}{c}\left(\frac{1}{\beta} - \beta\right) = \frac{\gamma d}{c}\ \frac{1-\beta^2}{\beta}$$

or: $\boxed{t_2 = \frac{d}{\gamma \beta c}}$ same as equation (2) obtained before.

****

8 A space ship is moving away from Earth with a speed $v = 0.8c$. When the ship is at a distance $d = 6.66 \times 10^8$ km from Earth, a radio signal is sent to the ship by the observers on Earth.
How long will it take for the signal to reach the ship:

(a) as measured by the scientists on Earth ?
(b) as measured by the crew of the space ship ?

***

Consider two inertial systems as shown, with their origins coinciding at the moment the space ship is at a distance d from Earth. The system S_1 is at rest with respect to Earth, while S_2 is at rest with respect to the space ship.

The Lorentz transformation between these two frames is:

$$x_2 = \gamma(x_1 - vt_1) \qquad (1)$$

$$t_2 = \gamma(t_1 - \frac{v}{c^2}x_1) \qquad (2)$$

In frame S_1, the signal is emitted at: $x_1 = -d, \quad t_1 = 0$

while in frame S_2 the emission of the signal has space-time coordinates, as per (1) and (2):

$$x_2 = \gamma(-d - v \times 0) = -\gamma d$$

$$t_2 = \gamma(0 + \frac{v}{c^2}d) = \frac{\gamma v d}{c^2}$$

The signal propagates in frame S_2 with speed c, and will reach the space ship (at $x_2 = 0$) after a time:

$$\Delta t_2 = \frac{\text{distance}}{\text{speed}} = \frac{\gamma d}{c} \qquad (3)$$

The event "arrival of the signal" will have in S_1 the coordinates:

$x_1' =$ location of ship @ $t_1' = vt_1' = \beta c t_1'$

$t_1' =$ the desired value

In frame S_2, using (1) and (2):

$$x_2' = \gamma(\beta c t_1' - \beta c t_1') = 0 \quad \text{, as expected,}$$

and:

$$t_2' = \gamma(t_1' - \frac{v}{c^2}\beta c t_1') = \gamma(1-\beta^2)t_1' = \frac{t_1'}{\gamma}$$

As seen by the observers in the space ship, then, the transit time of the radio signal is:

$$\Delta t_2 = t_2' - t_2 = \frac{t_1'}{\gamma} - \frac{\gamma v d}{c^2} \qquad (4)$$

Equating (3) and (4):

$$\frac{\gamma d}{c} = \frac{t_1'}{\gamma} - \frac{\gamma v d}{c^2}$$

and solving for t_1':

$$t_1' = \frac{\gamma^2 d}{c}(1+\beta)$$

$$\text{or:} \quad t_1' = \frac{d}{c(1-\beta)} \qquad (5)$$

Since $t_1 = 0$, this gives the transit time of the signal as seen by the observers on Earth, i.e., answer (a), while answer (b) is provided by (3).

Since all the information is given in frame of reference S_1, the transit time as measured by the scientists on earth can be obtained as follows. Let t_1' be the time when the signal reaches the ship. In that time, the radio signal has traveled a distance:

$$d_s = ct_1'$$

while the ship, that at $t_1' = 0$ was at a distance d, it is now at a distance:

$$d_{ss} = d + vt_1' = d + \beta ct_1'$$

Equating these two distances:

$$ct_1' = d + \beta ct_1'$$

and solving for t_1' :

$$t_1' = \frac{d}{c(1 - \beta)} \qquad \text{as before.}$$

<u>Numerically:</u> For $\beta = 0.8$, $\gamma = 5/3$.

Using (5): $$t_1' = \frac{6.66 \times 10^8 \times 10^5}{3 \times 10^{10} \times 0.20} = 1.11 \times 10^4 \text{ sec} \qquad \underline{\text{Ans.(a)}}$$

Using (3): $$\Delta t_2 = \frac{(5/3) \times 6.66 \times 10^8 \times 10^5}{3 \times 10^{10}} = 3700 \text{ sec} \qquad \underline{\text{Ans. (b)}}$$

Notice that the relation: $T = \gamma T_o$ cannot be applied directly, since it implies that the events separated by a time interval T_o occur at the same position (in frame S_2 in this case), which is not true.

Finally, at the moment the signal is received, the space ship is at a distance of the earth:

$$d_{ss} = d + \beta ct_1' = d(1 + \frac{\beta}{1-\beta}) = \frac{d}{1-\beta} = \frac{d}{0.2} = 5\ d$$

or: $$d_{ss} = 3.33 \times 10^9 \text{ km}$$

*****

9

The space ship *Antares* is traveling toward the star Tau Ceti and after 3 months is at a distance of 0.2 light-year as measured by observers on Earth. At that time, a second ship moving with speed 0.98c is sent after the *Antares*.

(a) How much later will the second ship overtake the *Antares*, as seen by the observers on Earth ?

(b) What will be the reading of the clocks on board the second ship when both ships meet ?

(c) At what distance from Earth, as seen by the observers on Earth, will the ships meet ?

***

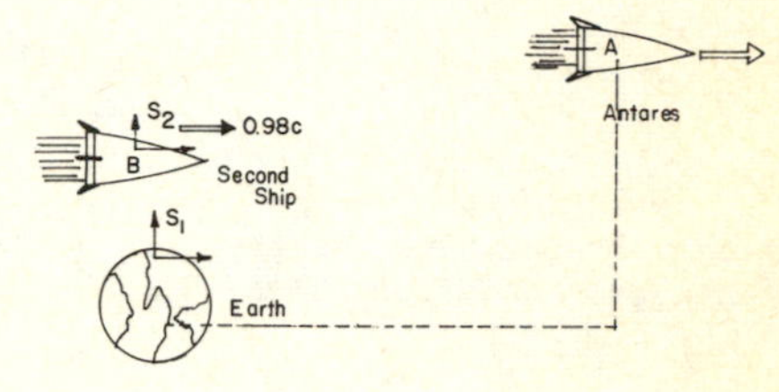

Consider two inertial frames: S_1, fixed w.r.t Earth, and S_2, fixed with respect to the second space ship, and thus moving w.r.t S_1 with speed 0.98c.
At $t_1 = t_2 = 0$, the origins coincide.
The speed of the *Antares* w.r.t Earth's observers is:

$$v = \frac{\text{distance}}{\text{time}} = \frac{0.2 \text{ ligh-year}}{0.25 \text{ year}} = 0.8c$$

Then: $\gamma = 5/3$. For the second ship: $\beta = 0.98$ c and: $\gamma = 5$.

Both ships move in the same frame of reference (S_1) and after some time t_1 will be at distances:

$$d_A = d_o + 0.8\, ct_1$$
$$d_B = 0.98\, ct_1$$

and will meet when: $d_A = d_B$, or: $0.8\, ct_1 + d_o = 0.98\, ct_1$

Solving for t_1: $$t_1 = \frac{d_o}{(0.98 - 0.8)c} = \frac{0.2 \text{ light-year}}{0.18 \times 1 \frac{\text{light-year}}{\text{year}}}$$

or: $$\boxed{t_1 = \frac{10}{9} = 1.11 \text{ years}}$$ Answer (a)

This result is verified by using the Lorentz transformation:

$$x_2 = \gamma(x_1 - vt_1) \qquad (1)$$
$$t_2 = \gamma(t_1 - \frac{v}{c^2}x_1) \qquad (2)$$
$$\dot{x}_2 = \frac{\dot{x}_1 - v}{1 - \frac{v}{c^2}\dot{x}_1} \qquad (3)$$

At $t_1 = 0$, the *Antares* is at $x_{1A} = 0.2$ light-year, so that its spacetime coordinates in inertial frame S_2 are, using (1) and (2):

$$x_{2A} = 5(0.2 - 0) = 1 \text{ ligh-year}$$
$$t_{2A} = 5(0 - 0.98 \times 0.2) = -0.98 \text{ years}$$

The velocity of the *Antares* in frame of reference S_2 is obtained from (3):

$$\dot{x}_2 = \frac{0.8 - 0.98}{1 - 0.98 \times 0.8} = - \frac{0.18}{1 - 0.784} = - 0.833 \frac{\text{light-year}}{\text{year}}$$

The negative sign indicates that the *Antares* is moving from right to left, toward the origin, $x_2 = 0$, where the second space ship is. Hence, both ships will meet after a time:

$$\Delta t = \frac{x_{2A}}{\dot{x}_{2A}} = \frac{1}{0.833} = 1.2 \text{ year}$$

and the clocks on board the second space ship will read:

$$t_2 = \text{initial time} + \Delta t = - 0.98 + 1.2$$

or:

$$\boxed{t_2 = 0.22 \text{ year}}$$

Answer (b)

This is proper time for the crew of the space ship, since the clocks remain at the same location. Thus, for the observers on Earth the elapsed time is:

$$t_1 = \gamma t_2 = 5 \times 0.22 = 1.1 \text{ years}, \quad \text{as before.}$$

In the frame of reference of Earth (S_1), at $t_1 = 0$ the *Antares* was at a distance

$$d_o = 0.2 \text{ light-year}$$

and moving with speed:

$$v = 0.8 \frac{\text{light-year}}{\text{year}}$$

Hence, at $t_1 = 10/9$ years it will be at:

$$d = d_o + vt_1 = 0.2 + 0.8 \times \frac{10}{9} = 1.09 \text{ light-year}$$

The second space ship, moving with speed 0.98 light-year/year will be at a distance:

$$d = vt_1 = 0.98 \times \frac{10}{9} = 1.09 \text{ light-year}$$

Thus, the ships meet at a distance from Earth:

$$\boxed{d = 1.09 \text{ light-year}}$$

Answer (c)

*****

10

A departing spacecraft drops a stick of length L nearby a space station S_o. The stick moves toward the station with constant velocity v_1 and remains parallel to the floor of S_o. The stick is observed by the crew of a second spaceship S_2, moving parallel to the space platform with constant velocity $\underline{v}_2$†.

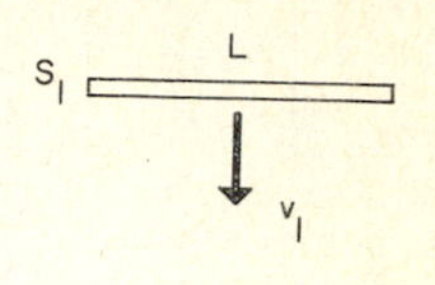

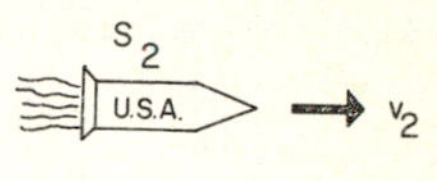

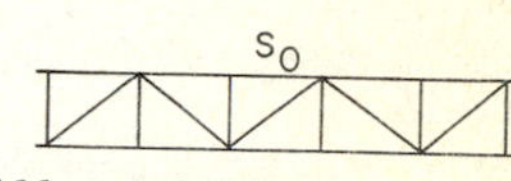

(a) Find formulas relating the coordinates (x_1, y_1, t_1) in system S_1 to the coordinates (x_2, y_2, t_2) in system S_2.

(b) Determine an expression for the angle with respect to the vertical at which the stick appears to drop for observer S_2.

(c) For observers S_o and S_1, all points of the stick will touch the space platform simultaneously, but for the observer S_2 the ends of L will hit the platform at different times. Calculate the difference between these times.

(d) The observer S_2 will see the stick at an angle with respect to the floor of the space station. Find an expression for that angle.

(e) Determine what the velocity $\underline{v}_1$ should be so that the stick would appear to observer S_2 as moving perpendicularly to itself.

****

he space station, the stick and the spaceship are respectively at rest in the ertial frames S_o, S_1, and S_2. The axes are parallel, and the +x coordinates e toward the right, while the +y coordinates are upward.

) The inertial frame S_1 is moving w.r.t. S_o with velocity $-v_1$ in the y-direction, or, S_o is moving w.r.t. S_1 with velocity v_1. Then, the Lorentz transformation written:

$$\left.\begin{aligned} x_o &= x_1 \\ y_o &= \gamma_1(y_1 - v_1 t_1) \\ t_o &= \gamma_1\left(t_1 - \frac{v_1}{c^2}\, y_1\right) \end{aligned}\right\} \qquad (1)$$

w, S_2 is moving w.r.t. S_o with velocity v_2 in the x-direction:

$$\left.\begin{aligned} x_2 &= \gamma_2(x_o - v_2 t_o) \\ y_2 &= y_o \\ t_2 &= \gamma_2\left(t_o - \frac{v_2}{c^2}\, x_o\right) \end{aligned}\right\} \quad \text{or, conversely:} \quad \left.\begin{aligned} x_o &= \gamma_2(x_2 + v_2 t_2) \\ y_o &= y_2 \\ t_o &= \gamma_2\left(t_2 + \frac{v_2}{c^2}\, x_2\right) \end{aligned}\right\} \qquad (2)$$

where:

$$\gamma_1 = \frac{1}{\sqrt{1-(v_1/c)^2}} = \frac{1}{\sqrt{1-\beta_1^2}}, \qquad \gamma_2 = \frac{1}{\sqrt{1-(v_2/c)^2}} = \frac{1}{\sqrt{1-\beta_2^2}}$$

S_o coordinates are now eliminated between (1) and (2):

This problem is closely related to the case, extensively treated in the literature, of a moving rod which is to pass through a slit in a wall parallel to it when slit is smaller than the length of the rod. See, for example, E.Marx, Am.J.Phys. 1127 (Dec.1967).

$$x_1 = \gamma_2(x_2 + v_2 t_2) \tag{3}$$

$$y_2 = \gamma_1(y_1 - v_1 t_1) \tag{4}$$

$$\gamma_1(t_1 - \frac{v_1}{c^2} y_1) = \gamma_2(t_2 + \frac{v_2}{c^2} x_2)$$

Solving the last equation for t_2:

$$t_2 = \frac{\gamma_1}{\gamma_2}(t_1 - \frac{v_1}{c^2} y_1) - \frac{v_2}{c^2} x_2 \tag{5}$$

Solving (3) for x_2:

$$x_2 = \frac{x_1}{\gamma_2} - v_2 t_2 \tag{6}$$

and replacing into (5):

$$t_2 = \frac{\gamma_1}{\gamma_2}(t_1 - \frac{v_1}{c^2} y_1) - \frac{v_2}{c^2} (\frac{x_1}{\gamma_2} - v_2 t_2)$$

or: $$t_2(1 - \frac{v_2^2}{c^2}) = \frac{\gamma_1}{\gamma_2} (t_1 - \frac{v_1}{c^2} y_1) - \frac{v_2}{c^2} \frac{x_1}{\gamma_2}$$

and, since $$1 - \frac{v_2^2}{c^2} = \frac{1}{\gamma_2^2}$$

finally: $$t_2 = \gamma_1 \gamma_2 (t_1 - \frac{v_1}{c^2} y_1) - \frac{v_2}{c^2} \gamma_2 x_1 \tag{7}$$

This relation is verified by noticing that if $v_1 = 0$, then $S_o \equiv S_1$, i.e., $\gamma_1 = 1$, $x_1 \equiv x_o$, $y_1 \equiv y_o$, $t_1 \equiv t_o$, and (7) becomes:

$$t_2 = \gamma_2(t_o - \frac{v_2}{c^2} x_o)$$

which is the Lorentz tranformation between frames S_2 and S_o. Or, if $v_2 = 0$, then S_2 is at rest in S_o, $\gamma_2 = 1$, and (7) becomes:

$$t_2 \equiv t_o = \gamma_1(t_1 - \frac{v_1}{c^2} y_1)$$

which is the third of equations (1).

Returning now to equation (6), one replaces t_2 by its expression (7):

$$x_2 = \frac{x_1}{\gamma_2} - v_2\gamma_1\gamma_2 t_1 + v_2 v_1 \frac{\gamma_1\gamma_2}{c^2} y_1 + \frac{v_2^2}{c^2} \gamma_2 x_1$$

$$= x_1(\frac{1}{\gamma_2} + \frac{v_2^2}{c^2} \gamma_2) - \gamma_1\gamma_2 v_2(t_1 - \frac{v_1}{c^2} y_1)$$

But:
$$\frac{1}{\gamma_2} + \frac{v_2^2}{c^2}\gamma_2 = \frac{1}{\gamma_2} + \beta_2^2\gamma_2 = \frac{1}{\gamma_2} + \frac{\gamma_2^2-1}{\gamma_2^2}\gamma_2 = \gamma_2$$

so, finally:

$$x_2 = \gamma_2 x_1 - \gamma_1\gamma_2 v_2\left(t_1 - \frac{v_1}{c^2}y_1\right)$$

Again, this expression is checked by making $v_1 = 0$. Then, $S_1 \equiv S_o$, $\gamma_1 = 1$, and:

$$x_2 = \gamma_2 x_o - \gamma_2 v_2 t_o = \gamma_2(x_o - v_2 t_o)$$

which is correct. Or, if $v_2 = 0$, then $S_2 \equiv S_o$, $\gamma_2 = 1$, and: $x_o = x_1$

The transformation of coordinates between frames S_1 and S_2 is expressed then:

$$\left.\begin{aligned} x_2 &= \gamma_2 x_1 - \gamma_1\gamma_2 v_2\left(t_1 - \frac{v_1}{c^2}y_1\right) \\ y_2 &= \gamma_1(y_1 - v_1 t_1) \\ t_2 &= \gamma_1\gamma_2\left(t_1 - \frac{v_1}{c^2}y_1 - \frac{v_2}{c^2}\frac{x_1}{\gamma_1}\right) \end{aligned}\right\} \quad (8)$$

The velocity transformations are now easily obtained. By differentiation of above equations one gets:

$$dx_2 = \gamma_2 dx_1 - \gamma_1\gamma_2 v_2\left(dt_1 - \frac{v_1}{c^2}dy_1\right)$$

$$dy_2 = \gamma_1(dy_1 - v_1 dt_1)$$

$$dt_2 = \gamma_1\gamma_2\left(dt_1 - \frac{v_1}{c^2}dy_1 - \frac{v_2}{c^2}\frac{1}{\gamma_1}dx_1\right)$$

hence:

$$\dot{x}_2 = \frac{dx_2}{dt_2} = \frac{\gamma_2 dx_1 - \gamma_1\gamma_2 v_2 dt_1 + (\gamma_1\gamma_2 v_1 v_2/c^2)dy_1}{\gamma_1\gamma_2 dt_1 - (\gamma_1\gamma_2 v_1/c^2)dy_1 - (\gamma_1\gamma_2 v_2/c^2\gamma_1)dx_1}$$

or:

$$\boxed{\dot{x}_2 = \frac{\dot{x}_1 - \gamma_1 v_2 + \dfrac{\gamma_1 v_1 v_2}{c^2}\dot{y}_1}{\gamma_1\left(1 - \dfrac{v_1}{c^2}\dot{y}_1\right) - \dfrac{v_2}{c^2}\dot{x}_1}} \quad (9)$$

Likewise:

$$\dot{y}_2 = \frac{dy_2}{dt_2} = \frac{\gamma_1 dy_1 - \gamma_1 v_1\, dt_1}{\gamma_1\gamma_2\, dt_1 - \dfrac{\gamma_1\gamma_2 v_1}{c^2}dy_1 - \dfrac{\gamma_1\gamma_2 v_2}{c^2\gamma_1}dx_1}$$

$$\text{or:} \qquad \boxed{\dot{y}_2 = \frac{\gamma_1(\dot{y}_1 - v_1)}{\gamma_1\gamma_2(1-\frac{v_1}{c^2}\dot{y}_1) - \frac{\gamma_2 v_2}{c^2}\dot{x}_1}} \qquad (10)$$

(b) The stick is at rest in the inertial frame S_1. Then, for any of its points:

$$\dot{x}_1 = \dot{y}_1 = 0$$

and equations (9) and (10) become:

$$\dot{x}_2 = - v_2 , \qquad \dot{y}_2 = - \frac{v_1}{\gamma_2}$$

which are the components of the velocity of the stick as seen by observer S_2. This velocity will form an angle θ with the vertical, such that:

$$\tan\theta = \frac{\dot{x}_2}{\dot{y}_2} = \frac{v_2}{v_1/\gamma_2}$$

$$\text{or:} \qquad \boxed{\tan\theta = \frac{v_2\gamma_2}{v_1}}$$

(c) The coordinates of the ends of the stick, in its own frame of reference, are:

$$\text{point A:} \begin{cases} x_1^A = 0 \\ y_1^A = 0 \end{cases} \qquad \text{point B:} \begin{cases} x_1^B = L \\ y_1^B = 0 \end{cases}$$

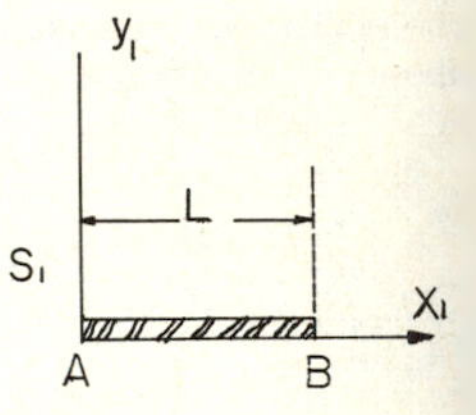

Hence, for observer S_2, using equations (8), one has:

$$\text{point A:} \begin{cases} x_2^A = - \gamma_1\gamma_2 v_2 t_1^A \\ y_2^A = - \gamma_1 v_1 t_1^A \\ t_2^A = \gamma_1\gamma_2 t_1^A \end{cases} \qquad (11)$$

$$\text{point B:} \begin{cases} x_2^B = \gamma_2 L - \gamma_1\gamma_2 v_2 t_1^B \\ y_2^B = - \gamma_1 v_1 t_1^B \\ t_2^B = \gamma_1\gamma_2 t_1^B - \gamma_2 \frac{v_2}{c^2} L \end{cases} \qquad (12)$$

Now, for observer S_1, both ends of the stick reach the space platform at the same time, say:

$$t_1 = t_1^A = t_1^B$$

while for observer S_2, using above equations:

$$t_2^A = \gamma_1\gamma_2 t_1$$
$$t_2^B = \gamma_1\gamma_2 t_1 - \gamma_2 \frac{v_2}{c^2} L$$

Thus: $$\boxed{\Delta t_2 = \gamma_2 \frac{v_2}{c^2} L}$$ Ans. (c)

Notice that $\Delta t_2 = 0$ only if $v_2 = 0$, i.e., if the space ship is on the space platform.

(d) After the end B of the stick hits the platform, a certain time Δt_2 must elapse before the end A will hit. But, for observer S_2, all the points of the stick have the same vertical velocity, as obtained in (b) above:

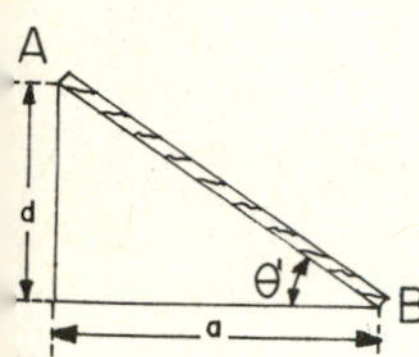

$$\dot{y}_2 = -\frac{v_1}{\gamma_2}$$

Hence: $$d = |\dot{y}_2|\ \Delta t_2 = \frac{v_1}{\gamma_2}\frac{\gamma_2 v_2}{c^2} L = \frac{v_1 v_2}{c^2} L$$

At any given time, say: $t_2 = t_2^B = t_2^A$, the distance a is, for observer S_2, given by the difference:

$$x_2^B - x_2^A$$

which is obtained using equations (11) and (12) as follows:

$$x_2^A = -\gamma_1\gamma_2 v_2 t_1^A = -v_2(\gamma_1\gamma_2 t_1^A) = -v_2 t_2^A$$
$$x_2^B = \gamma_2 L - v_2(\gamma_1\gamma_2 t_1^B) = \gamma_2 L - v_2(t_2^B + \gamma_2 \frac{v_2}{c^2} L)$$

Thus: $$a = x_2^B - x_2^A = \gamma_2 L(1 - \frac{v_2^2}{c^2})$$

or: $$a = \frac{L}{\gamma_2}$$ as expected.

The stick will appear to observer S_2 as tilted with an angle θ' given by:

$$\tan\theta' = \frac{d}{a} = \frac{v_1 v_2}{c^2} L \frac{\gamma_2}{L}$$

or: $$\boxed{\tan\theta' = \gamma_2 \frac{v_1 v_2}{c^2}} \quad (13)$$ Ans. (d)

Discussion:

Notice that θ' = 0 only in two cases:

(i) $v_1 = 0$, i.e., the stick is not dropping

and: (ii) $v_2 = 0$, i.e., the space ship has landed on the platform.

Moreover, let: tan θ' = k. Then, equation (13) becomes:

$$k^2 = \frac{\beta_1^2\beta_2^2}{1-\beta_2^2}$$

and solving for β_1^2:

$$\beta_1^2 = \frac{k^2(1-\beta_2^2)}{\beta_2^2}$$

showing that the same angle θ' can be obtained for many combinations of the velocities v_1 and v_2.

(e) The observer S_2 will see the stick moving perpendicularly to itself when:

$$\tan\theta = \tan\theta'$$

or:

$$\frac{v_2\gamma_2}{v_1} = \frac{\gamma_2 v_1 v_2}{c^2}$$

which simplifies to: $v_1^2 = c^2$ or: $\boxed{v_1 = c}$

corresponding to the case when the stick is moving with the speed of light. Notice that this result is independent of the speed of the space ship ! Since we have divided by v_2, a second solution is $v_2 = 0$, i.e., when the space ship is resting on the space station.

****

11 A particle accelerator supplies relativistic electrons having momentum $p = 1.1 \times 10^{-21}$ newton sec. Calculate the velocity, the mass and the kinetic energy in Mev of the electrons.

Answers: $v = 2.913 \times 10^{8}$ m/sec, $m = 3.78 \times 10^{-30}$ Kg, $K = 1.61$ Mev.

****

12 Consider an electron having kinetic energy equal to its rest energy. Show that the energy of a photon which has the same momentum as this electron is given by:

$$E_\gamma = \sqrt{3}\, E_o$$

where: E_o = rest energy of the electron.

****

13 A space ship is approaching Earth with a speed $v = 0.8c$. When the ship is at a distance $d = 6.66 \times 10^{8}$ km from Earth, a radio signal is sent to the ship by the observers on Earth.
How long will it take for the signal to reach the ship:

(a) as measured by the scientists on Earth ?
(b) as measured by the crew of the space ship ?

Answers: (a) 1235 sec, (b) 3700 sec.

****

2

INTERACTION OF RADIATION AND MATTER

14 What is the kinetic energy of the photoelectrons ejected from a tungsten surface by ultraviolet light of wavelength 1940 Å ? Express the result in ev. The photoelectric threshold of tungsten is 2300 Å.

****

The kinetic energy of the photoelectrons is given by the equation:

$$K = \frac{1}{2} mv^2 = h(\nu - \nu_o) = hc(\frac{1}{\lambda} - \frac{1}{\lambda_o})$$

or:

$$K = \frac{hc}{\lambda\lambda_o} (\lambda_o - \lambda)$$

By numerical substitution, and recalling that 1 ev = 1.6×10^{-19} joule, one has:

$$K = \frac{6.626 \times 10^{-34} (\text{joule-sec}) \times 3 \times 10^{8} (\text{m/sec}) \times (2.3 - 1.94) \times 10^{-7} (\text{m})}{1.94 \times 10^{-7} (\text{m}) \times 2.30 \times 10^{-7} (\text{m}) \times 1.6 \times 10^{-19} (\text{joule/ev})}$$

or: $\boxed{K = 1 \text{ ev}}$ Answer

*****

15

The work function of sodium is 2ev.
(a) Find the maximum energy of the photoelectrons ejected when a sodium surface is illuminated with light of 3105 Å.
(b) Calculate the maximum wavelength of the light that would produce photoelectric effect.

****

(a) The basic formula for photoelectric effect is:

$$\boxed{h\nu = \frac{1}{2} mv^2 + W}$$

where W is the work function. Then:

$$K = \frac{1}{2} mv^2 = h\nu - W \qquad (1)$$

For λ = 3105Å the energy of the photons is:

$$E_\gamma = h\nu = \frac{hc}{\lambda} = \frac{6.626 \times 10^{-34} \times 3 \times 10^{8}}{3105 \times 10^{-10} \times 1.6 \times 10^{-19}} \text{ ev}$$

or: $E_\gamma = 4$ ev

Hence, in (1): $K = 4 - 2 = 2$ or: $\boxed{K = 2 \text{ ev}}$ Ans.(a)

(b) The threshold energy of the photons is:

$$h\nu_o = W$$

Then:

$$\lambda_o = \frac{hc}{W} = \frac{6.626 \times 10^{-34} \times 3 \times 10^{8}}{2 \times 1.6 \times 10^{-19}} \times 10^{10} \text{ Å}$$

or: $\boxed{\lambda_o = 6210 \text{ Å}}$ Ans.(b)

****

16

The work function of cadmium is 4.07 ev. (a) Determine the threshold wavelength for the emission of photoelectrons; (b) what would be the wavelength of the incident radiation if the photoelectrons are to be ejected with velocity $v = 0.1c$? (c) what wavelength is required for the electrons to have $v = 0.866c$? What is the energy of the photons in this case ?

****

(a) Einstein's equation for the photoelectric effect is:

$$\boxed{K = h\nu - W} \qquad (1)$$

At the threshold frequency, $K = 0$ and the photon's energy is just equal to the work function W:

$$h\nu_o = \frac{hc}{\lambda_o} = W$$

Then:

$$\lambda_o = \frac{hc}{W} = \frac{6.626 \times 10^{-27}(\text{erg-sec}) \times 3 \times 10^{10}(\text{cm/sec})}{4.07\ (\text{ev}) \times 1.6 \times 10^{-12}\ (\text{erg/ev})} = 3.052 \times 10^{-5}\ \text{cm}$$

or:

$$\boxed{\lambda_o = 3052\ \text{Å}} \qquad \underline{\text{Ans.(a)}}$$

(b) Using the classical expression for the kinetic energy:

$$K = \frac{1}{2}mv^2 = \frac{1}{2}mc^2\ \beta^2 = \frac{1}{2} \times 0.511 \times 0.01 = 2.55 \times 10^{-3}\ \text{Mev}$$

or: $K = 2550$ ev

Since $v = 0.1c$ is the breaking point at which relativistic effects become important, one verifies the above result by using the correct relativistic expression for the kinetic energy:

$$K = m_oc^2(\gamma-1) \qquad \text{where, for } \beta = 0.1,\quad \gamma = 1.005.$$

Then: $K = 0.511 \times (1.005 - 1) = 0.511 \times 0.005 = 2.555 \times 10^{-3}$ as above.

Equation (1) is written:

$$\frac{hc}{\lambda} = K + W = 2550 + 4.07 = 2554\ \text{ev} \qquad (2)$$

or:

$$\lambda = \frac{hc}{2554} = \frac{6.626 \times 10^{-27} \times 3 \times 10^{10}}{2554 \times 1.6 \times 10^{-12}} = 4.86 \times 10^{-8}\ \text{cm}$$

$$\boxed{\lambda = 4.86\ \text{Å}} \qquad \underline{\text{Ans.(b)}}$$

(c) For $v = 0.866c$, the photoelectrons are relativistic. Since to $\beta = 0.866$ corresponds $\gamma = 2$, the kinetic energy is:

$$K = m_oc^2(\gamma-1) = 0.511 \times (2-1) = 0.511\ \text{Mev}$$

and using (2):

$$\frac{hc}{\lambda} = 0.511 \times 10^6 + 4.07 \simeq 0.511 \times 10^6\ \text{ev}$$

Then: $$\lambda = \frac{hc}{K+W} = \frac{6.626 \times 10^{-27} \times 3 \times 10^{10}}{0.511 \times 10^{6} \times 1.6 \times 10^{-12}} = 2.43 \times 10^{-10} \text{ cm}$$

or: $$\boxed{\lambda = 0.0243 \text{ Å}}$$ Ans. (c)

Since W is negligible compared to the kinetic energy of the electron, the energy of the incident photon is:

$$\boxed{h\nu = 0.511 \text{ Mev}}$$ Ans.(c)

Notice, however, that the probability for photoelectric effect to occur decreases rapidly with increasing frequency of the incident photon. Yet, in certain cases, the fact that the ejected electron could be relativistic must be taken into consideration.

****

17 Prove that a photon cannot transfer all its energy to an isolated electron.

****

Since the momentum is in the direction of the incident photon, after the interaction the electron will be moving also in that direction, as shown in the figure.

$h\nu$ m_o m v

before after

By conservation of energy:

$$h\nu + m_oc^2 = mc^2 \qquad (1)$$

where m_o, the rest mass of the electron, and m, its relativistic mass, are related by the expression:

$$m = \gamma m_o$$

Here, as usual: $\gamma = 1/\sqrt{1-\beta^2}$ and: $\beta = v/c$.

Equation (1) is written then:

$$h\nu = m_oc^2(\gamma-1) \qquad (2)$$

By conservation of momentum: $\frac{h\nu}{c} = mv = \gamma m_o\beta c$

$$\text{or:} \qquad h\nu = \gamma\beta m_oc^2 \qquad (3)$$

Equating (2) and (3): $m_oc^2(\gamma-1) = \gamma\beta m_oc^2$

$$\text{or:} \qquad \gamma - 1 = \gamma\beta$$

$$\text{or also:} \qquad \frac{1}{\sqrt{1-\beta^2}} - 1 = \frac{\beta}{\sqrt{1-\beta^2}}$$

From here: $1 - \sqrt{1-\beta^2} = \beta$ or: $1 - \beta = \sqrt{1-\beta^2}$

and, finally: $\boxed{\beta(1-\beta) = 0}$

Hence, either $\beta = 0$, and the electron is at rest after the interaction, or $\beta = 1$, i.e., the electron is moving with the speed of light. Both cases are impossible. Consequently, simultaneous conservation of energy and momentum requires the presence of a third body (namely, the nucleus).

*****

18 In a Compton scattering experiment the difference in frequency bewteen the primary photon and the scattered photon is 1.233×10^{20} cycles/sec. What is the speed of the recoil electron ?

****

By conservation of energy:

$$h\nu = h\nu' + K$$

or: $$K = h(\nu-\nu') = h\,\Delta\nu$$

Numerically:

$$K = \frac{6.626 \times 10^{-27} \times 1.233 \times 10^{20}\ \text{(erg)}}{1.6 \times 10^{-6}\ \text{(erg/Mev)}} = 0.511\ \text{Mev}$$

i.e., the kinetic energy of the recoil electron is equal to its rest mass. Then, since:

$$K = E_o(\gamma-1)$$

it follows: $\gamma = 2$, and from tables: $\beta = 0.866$.

The speed of the recoil electron is then:

$$v = \beta c = 0.866 \times 3 \times 10^{10}\ \text{cm/sec}$$

or: $$\boxed{v = 2.6 \times 10^{10}\ \text{cm/sec}}$$ Answer

****

19 Prove the following equations for the Compton effect:

$$\lambda' = \lambda + \frac{h}{m_o c}(1-\cos\theta) \qquad \text{(a)}$$

$$h\nu' = \frac{h\nu}{1+\alpha(1-\cos\theta)} \qquad \text{(b)}$$

$$\cot\frac{\theta}{2} = (1+\alpha)\tan\phi \qquad \text{(c)}$$

where: $\alpha = h\nu/m_o c^2$

θ = angle of scattered photon

ϕ = angle of recoil electron

****

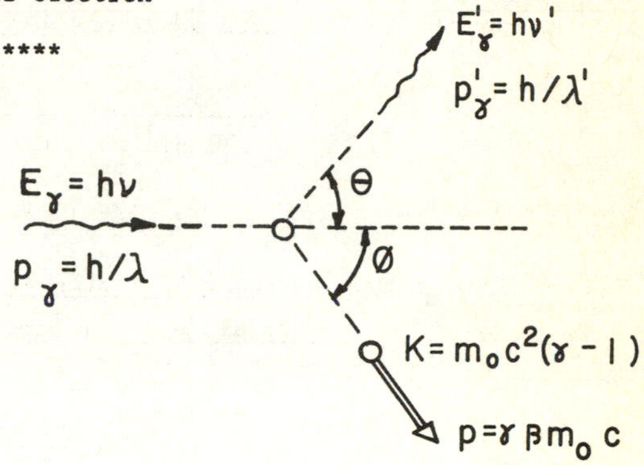

Using the notation indicated in the figure, conservation of energy is written:

$$h\nu + m_o c^2 = h\nu' + \gamma m_o c^2 \qquad (1)$$

and conservation of momentum:

$$\frac{h}{\lambda} = \frac{h}{\lambda'}\cos\theta + \gamma\beta m_o c\cos\phi \qquad (2)$$

$$0 = \frac{h}{\lambda'}\sin\theta - \gamma\beta m_o c\sin\phi \qquad (3)$$

Squaring and adding equations (2) and (3):

$$\frac{h^2}{\lambda^2} + \frac{h^2}{\lambda'^2} - \frac{2h^2}{\lambda\lambda'}\cos\theta = \gamma^2\beta^2 m_o^2 c^2 = \gamma^2\,\frac{\gamma^2-1}{\gamma^2}\,m_o^2 c^2 = \gamma^2 m_o^2 c^2 - m_o^2 c^2 \qquad (4)$$

The term $\gamma^2 m_o^2 c^2$ is obtained from (1):

$$\gamma^2 m_o^2 c^2 = \frac{1}{c^2}(h\nu + m_o c^2 - h\nu')^2$$

and by substitution into (4) one gets:

$$\frac{h^2}{\lambda^2} + \frac{h^2}{\lambda'^2} - \frac{2h^2}{\lambda\lambda'}\cos\theta = \frac{1}{c^2}\left[\frac{h^2c^2}{\lambda^2} + \frac{h^2c^2}{\lambda'^2} + m_o^2c^4 - \frac{2h^2c^2}{\lambda\lambda'} + \frac{2hc}{\lambda}m_o c^2 - \frac{2hc}{\lambda'}m_o c^2\right] - m_o^2 c^2$$

which after simplification yields:

$$\boxed{\lambda' - \lambda = \frac{h}{m_o c}(1-\cos\theta)} \qquad (5)$$

which is equation (a) for the wavelength shift. Here,

$$\frac{h}{m_o c} = \text{Compton wavelength} = 0.02426\ \text{Å}$$

Inverting equation (5):

$$\frac{1}{\lambda'} = \frac{1}{\lambda + \frac{h}{m_o c}(1 - \cos\theta)} = \frac{1/\lambda}{1 + \frac{h\nu}{m_o c^2}(1 - \cos\theta)}$$

and multipliying by hc:

$$\boxed{h\nu' = \frac{h\nu}{1 + \alpha(1-\cos\theta)}} \qquad (6)$$

which is equation (b), Q.E.D.

Dividing now the two momentum equations (2) and (3):

$$\cot\phi = \frac{\frac{h}{\lambda} - \frac{h}{\lambda'}\cos\theta}{\frac{h}{\lambda'}\sin\theta} = \frac{h\nu - h\nu'\cos\theta}{h\nu'\sin\theta}$$

Using equation (6):

$$\cot\phi = \frac{h\nu - h\nu\cos\theta/[1 + \alpha(1-\cos\theta)]}{\frac{h\nu\sin\theta}{1 + \alpha(1-\cos\theta)}} = \frac{1 + \alpha(1-\cos\theta) - \cos\theta}{\sin\theta}$$

or:

$$\cot\phi = \frac{1 - \cos\theta}{\sin\theta}(1+\alpha) \qquad (7)$$

Recalling the trigonometric identity:

$$\tan\frac{\theta}{2} = \frac{1 - \cos\theta}{\sin\theta}$$

equation (7) becomes:

$$\cot\phi = (1+\alpha)\tan\frac{\theta}{2}$$

and finally:

$$\boxed{\cot\frac{\theta}{2} = (1+\alpha)\tan\phi} \qquad \text{Q.E.D.}$$

****

20 Show that the energy of the radiation scattered at 90° by Compton effect cannot exceed the rest energy of the electron, m_oc^2.

****

The Compton shift is expressed:

$$\boxed{\lambda' - \lambda = \frac{h}{m_oc}(1 - \cos\theta)}$$

which, for $\theta = 90°$, becomes:

$$\lambda' - \lambda = \frac{h}{m_oc}$$

If the energy of the incident photon increases, its frequency increases and λ decreases, until it can be neglected when compared to λ'. Then:

$$\lambda' = \frac{h}{m_oc}$$

and the energy of the scattered photon is:

$$E'_\gamma = h\nu' = \frac{hc}{\lambda'} = \frac{hc}{h/m_oc}$$

or: $$\boxed{E'_\gamma = m_oc^2}$$ Q.E.D.

****

21

In a Compton collision the scattered photon is observed to move at 90° with respect to the direction of the initial photon. If the energy of the incident photon is 3 Mev, calculate the energy of the scattered photon and the kinetic energy of the recoil electron, both in terms of Mev.

****

The basic relation for Compton scattering is:

$$\boxed{\lambda' - \lambda = \frac{h}{m_o c}(1 - \cos\theta)} \qquad (1)$$

where: λ = wavelength of incident photon

λ' = wavelength of scattered photon

θ = angle of scattered photon = 90°, $\cos\theta = 0$.

Since: $\lambda = \frac{c}{\nu} = \frac{hc}{h\nu} = \frac{hc}{E_\gamma}$, then: $\frac{\lambda}{hc} = \frac{1}{E_\gamma}$, $\frac{\lambda'}{hc} = \frac{1}{E'_\gamma}$

and equation (1) becomes: $\frac{1}{E'_\gamma} = \frac{1}{E_\gamma} + \frac{1}{m_o c^2}$

Numerically:

$$\frac{1}{E'_\gamma} = \frac{1}{3} + \frac{1}{0.511} = 0.333 + 1.96 = 2.293$$

and: $\boxed{E'_\gamma = 0.436 \text{ Mev}}$

Now, by conservation of energy:

$$E_\gamma = E'_\gamma + K$$

and the kinetic energy of the recoil electron is:

$$K = E_\gamma - E'_\gamma = 3.000 - 0.436$$

or: $\boxed{K = 2.564 \text{ Mev}}$

****

22

A beam of 40 Kev X-rays is incident upon an aluminum foil.
(a) Determine the maximum and minimum energy of the scattered photons.
(b) What is the maximum energy of the recoil electrons ?

****

(a) For Compton effect, the energy of the scattered photons is given by:

$$\boxed{h\nu' = \frac{h\nu}{1 + \alpha(1-\cos\theta)}} \qquad (1)$$

where: $\alpha = \dfrac{h\nu}{m_o c^2}$

θ = angle between the directions of incident and scattered photon.

The maximum energy of the scattered photon is obtained when the denominator of (1) is minimum, i.e., when $\cos\theta = 1$, or $\theta = 0$. Thus, $h\nu' = h\nu$, which physically corresponds to the case when the incident photon suffers no interaction. Then:

$$\boxed{\max h\nu' = 40 \text{ Kev}}$$ Answer

The minimum energy corresponds to the maximum of the denominator, i.e., when $\cos\theta = -1$, or $\theta = 180°$. The ejected photon is directed backwards, and will have an energy:

$$h\nu' = \frac{h\nu}{1+2\alpha}$$

Numerically:

$$\left.\begin{array}{l} h\nu = 40 \text{ Kev} \\ \alpha = \dfrac{40}{511} = 0.0783 \end{array}\right\} \qquad h\nu' = \frac{40}{1 + 2 \times 0.0783} = 34.6 \text{ Kev}$$

or: $$\boxed{\min h\nu' = 34.6 \text{ Kev}}$$ Answer

(b) By conservation of energy, the recoil electron will have a kinetic energy:

$$K = h\nu - h\nu'$$

which will be maximum when $h\nu'$ is minimum, i.e., for $h\nu' = 34.6$ Kev.

Hence: $$K = 40 - 34.6 = \boxed{5.4 \text{ Kev}}$$ Answer

****

23 Show that pair production cannot occur for an isolated photon.

****

Consider the most general case, when the particles are ejected in arbitrary directions θ and $\emptyset$. Let E^+ and E^- be their total energies. By conservation of energy:

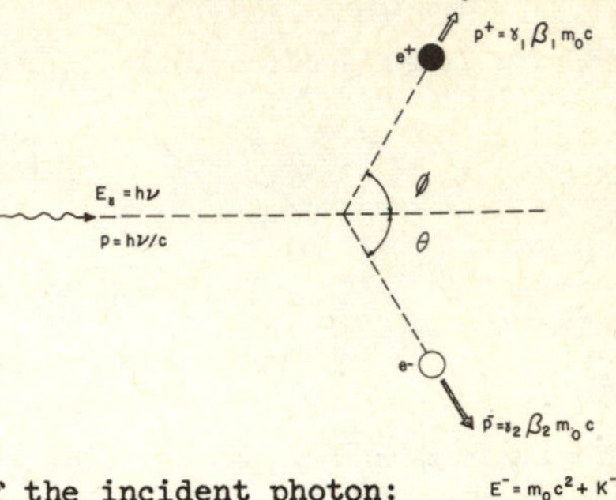

$$h\nu = 2m_oc^2 + m_oc^2(\gamma_1 - 1) + m_oc^2(\gamma_2 - 1)$$

$$\text{or:} \qquad h\nu = m_oc^2(\gamma_1 + \gamma_2) \qquad (1)$$

where, as usual: $\gamma_i = \dfrac{1}{\sqrt{1-\beta_i^2}}$, i =1, 2.

By conservation of momentum in the direction of the incident photon:

$$\frac{h\nu}{c} = \gamma_1 m_o \beta_1 c \cos \emptyset + \gamma_2 m_o \gamma_2 c \cos \theta$$

$$\text{or:} \qquad h\nu = m_o c^2 (\gamma_1 \beta_1 \cos \emptyset + \gamma_2 \beta_2 \cos \theta) \qquad (2)$$

Simultaneous verification of (1) and (2) requires that:

$$\beta_1 \cos \emptyset = 1, \qquad \text{and:} \qquad \beta_2 \cos \theta = 1$$

which are also written:

$$\cos \emptyset = \frac{1}{\beta_1} \leqslant 1 , \qquad \text{and:} \qquad \cos \theta = \frac{1}{\beta_2} \leqslant 1$$

$$\text{i.e.,} \qquad \beta_1 \geqslant 1 , \text{ and:} \quad \beta_2 \geqslant 1$$

Since $\beta_i = v_i/c$, i=1,2, this is not possible. Thus, conservation of energy and momentum are verified only if a third body is present, as for example, the nucleus.

*****

24

Prove that the minimum photon energy required to produce an electron-positron pair in the neighborhood of an electron is $4m_oc^2$. Show that in that case the three particles are moving in the LCS with kinetic energy $K = 2m_oc^2/3$ each†.

****

In general, the energy of the incident photon will be spent on creating the pair and in given kinetic energy to the existing electron and to the new particles. The minimum energy required will then correspond to the case when the three particles are at rest in the center-of-mass coordinate system (CMCS). But then, in the laboratory system (LCS), the three particles will be moving with the same velocity, as shown in the figure.

LCS

before $h\nu$ m_oc^2

after

Let K_i represent the kinetic energy of the particles. Then:

$$K_1 = K_2 = K_3 = K$$

and conservation of energy is expressed:

$$h\nu + m_oc^2 = K_1 + K_2 + K_3 + 3m_oc^2$$

or:

$$h\nu = 3K + 2m_oc^2 \tag{1}$$

By conservation of momentum:

$$\frac{h\nu}{c} = 3p = \frac{3}{c}\sqrt{K(K + 2m_oc^2)} \tag{2}$$

where the relativistic expression for the momentum of the particles has been used. From (1) and (2):

$$3K + 2m_oc^2 = 3\sqrt{K(K + 2m_oc^2)}$$

Squaring:

$$9K^2 + 4(m_oc^2)^2 + 12K\, m_oc^2 = 9K^2 + 18K\, m_oc^2$$

or:

$$\boxed{K = \frac{2}{3} m_oc^2} \qquad \text{Q.E.D.}$$

Then, by substitution into equation (1):

$$h\nu = 3\,\frac{2}{3}\, m_oc^2 + 2m_oc^2$$

or:

$$\boxed{h\nu = 4\, m_oc^2} \qquad \text{Q.E.D.}$$

*****

†) This problem has been extensively treated in the literature, often in a more complicated manner. See, for example: (i) Perrin, Compt.rend., 197, 1100 (Nov.13, 1933);(ii) K. M. Watson, Phys. Rev.72 (1947); (iii) A.Borsellino, Revista de la Universidad Nacional de Tucumán, Vol.6, No.1 (Dec.1947).

25

Prove that an isolated charged particle in vacuum cannot radiate a photon.

****

Consider the charged particle at rest, i.e., the problem will be solved in the CMCS. Let m_o be the rest mass of the charged particle, E its total energy and p its momentum.

before: m_oc^2

after: E,P; $m=\gamma m_o$; $E_\gamma = h\nu$, $p = h/\lambda$

By conservation of energy:

$$m_oc^2 = E + h\nu \qquad (1)$$

where: $$E^2 = E_o^2 + p^2c^2 = (m_oc^2)^2 + p^2c^2 \qquad (2)$$

By conservation of momentum:

$$\frac{h\nu}{c} = p, \qquad \text{or:} \quad h\nu = pc, \qquad \text{or:} \quad h^2\nu^2 = p^2c^2 \qquad (3)$$

Solving (1) for E and squaring:

$$E = m_oc^2 - h\nu$$

$$E^2 = (m_oc^2)^2 + h^2\nu^2 - 2m_oc^2\,h\nu$$

and equating to (2):

$$p^2c^2 + m_o^2c^4 = m_o^2c^4 + h^2\nu^2 - 2m_oc^2\,h\nu$$

Using (3): $$h^2\nu^2 = h^2\nu^2 - 2m_oc^2\,h\nu$$

which is only satisfied if $\nu = 0$, i.e., if no photon is radiated.

****

26 The intensity of full sunlight at the earth's surface is estimated to be 0.13 watts/cm^2.
(a) Calculate the radiation pressure on a perfectly reflecting surface placed normally to the direction of the light.
(b) Find the amplitude of the electric field strength of this beam. Express the result in volts/meter.

****

(a) Since photons have no rest mass: $E_\gamma = pc$, or: $p = \frac{E_\gamma}{c}$

Now, by Newton's second law: $F = \frac{\Delta p}{\Delta t}$

and for complete reflection the change in momentum is:

$$\Delta p = p - (-p) = 2p$$

The force is then:
$$F = \frac{\Delta p}{\Delta t} = \frac{2p}{\Delta t} = \frac{2E_\gamma}{c\Delta t}$$

and the pressure:
$$P = \frac{F}{\Delta A} = \frac{2E_\gamma}{c\ \Delta t \Delta A}$$

But $E_\gamma/\Delta t\Delta A$ is the energy deposited by the radiation per unit area per unit time, i.e., the intensity of the radiation I. Thus:

$$\boxed{P = \frac{2I}{c}} \qquad (1)$$

Notice that for the case when the surface is a perfect absorber, the change of momentum would be:

$$\Delta p = p - 0 = p$$

and expression (1) would be:
$$\boxed{P = \frac{I}{c}} \qquad (2)$$

Numerically:

$$P = \frac{2 \times 0.13\ (\text{watt/cm}^2)}{3 \times 10^{10}\ (\text{cm/sec})} \times 10^7 \frac{(\text{erg/sec})}{(\text{watt})}$$

or:
$$\boxed{P = 8.66 \times 10^{-5}\ \text{dyne/cm}^2}$$
Ans.(a)

b) The electric field strength and the intensity of the radiation are related by the expression:

$$I = \varepsilon_o E^2 c$$

valid in the MKS system of units. Here:

$$\varepsilon_o = \text{permittivity} = \frac{1}{36\pi \times 10^9}\ \frac{\text{farad}}{\text{m}}$$

$$c = 3 \times 10^8\ \text{m/sec}$$

Then:
$$E^2 = \frac{I}{\varepsilon_o c} = \frac{0.13 \times 10^4 (\text{watt/m}^2) \times 36\pi \times 10^9\ (\text{m/farad})}{3 \times 10^8\ (\text{m/sec})} = 49 \times 10^4\ (\text{volt/m})^2$$

or:
$$\boxed{E = 700\ \text{volts/m}}$$
Ans.(b)

****

27 Moonlight has at the earth's surface a maximum intensity of 0.82×10^{-3} watts/m^2.

(a) Calculate the force due to the moonlight on a perfectly absorbing surface of area 100 m^2, when the moon is exactly overhead.

(b) If the average wavelength of the moonlight is 5000 Å, what is the number of photons/cm^3 in the beam ?

(c) Calculate the amplitude of the electric field strength of the beam. Do this using both the MKS and CGS systemsof units. Express the result in volts/meter.

****

(a) The radiation pressure is given by the formula developed in the previous problem:

$$\boxed{P = \frac{I}{c}}$$

and the force on an area A is: $F = PA = \frac{IA}{c}$

Numerically:

$$F = \frac{0.82 \times 10^{-3} (\text{watt/m}^2) \times 100 \text{ m}^2}{3 \times 10^8 \text{ (m/sec)}}$$

or: $\boxed{F = 0.273 \times 10^{-9} \text{ newton}}$ Ans.(a)

(b) Let: n = number of photons per unit volume of beam

N = photon flux, i.e., photons per unit area per unit time

$h\nu$ = energy of each photon

Then:

$$I = N h\nu \qquad (1)$$

Consider an area unity perpendicular to the beam. Since the beam is moving with speed c, all photons within a distance c of that unit area will cross it in one second. Thus:

$$N = nc \ , \quad \text{or:} \quad n = \frac{N}{c}$$

and combining with (1):

$$n = \frac{I}{ch\nu} = \frac{I\lambda}{hc^2}$$

By numerical substitution, using MKS units:

$$n = \frac{0.82 \times 10^{-3} (\text{watt/m}^2) \times 5000 \ (\text{Å}) \times 10^{-10} (\text{m/Å})}{6.626 \times 10^{-34} (\text{joule-sec}) \times 9 \times 10^{16} (\text{m}^2/\text{sec}^2)}$$

or: $\boxed{n = 6.87 \times 10^6 \text{ photons/m}^3 = 6.87 \text{ photons/cm}^3}$ Ans.(b)

(c) Using MKS units, the intensity of the radiation and the electric field strength E are related by the equation:

$$\boxed{I = \varepsilon_o E^2 c}$$

where: ε_o = permittivity = $1/36\pi \times 10^9$ farad/m ($\equiv$ coul^2/newton-m^2)

Then: $$E^2 = \frac{0.82 \times 10^{-3} \times 36\pi \times 10^9}{3 \times 10^8} = 0.309\ (\text{volt/m})^2$$

and: $$\boxed{E = .556 \text{ volts/m}}$$ Ans.(c)

If one uses now the CGS system of units, by definition $\varepsilon_o = 1$, and intensity and electric field strength are related as follows:

$$\boxed{I = \frac{\varepsilon_o E^2 c}{4\pi}}$$

Thus:

$$E^2 = \frac{4\pi I}{c} = \frac{4\pi \times 0.82}{3 \times 10^{10}} = 3.435 \times 10^{-10}\ (\text{statvolt/cm})^2$$

$$\therefore \qquad E = 1.853 \times 10^{-5} \text{ statvolts/cm}$$

But: $$\left.\begin{array}{r} 1 \text{ statvolt} = 300 \text{ volts} \\ 1 \text{ cm} = 10^{-2} \text{ meter} \end{array}\right\} \therefore \quad 1 \text{ statvolt/cm} = 3 \times 10^4 \text{ volts/m}$$

Hence:

$$E = 1.853 \times 10^{-5} \times 3 \times 10^4 \text{ volts/m}$$

or $$\boxed{E = 0.556 \text{ volts/m}}$$ as before.

****

28

A 8-watt beam of light strikes a surface normally. The surface reflects 50% of the incident radiation, the remainder being absorbed. Determine the force exerted on the surface by the radiation.
If the average wavelength is 5000 Å, and the beam has a cross section of 10 cm^2, calculate the number of photons per cm^3 in the beam.

****

By Newton' second law: $F = \frac{\Delta p}{\Delta t} = \frac{ap}{\Delta t}$ where: $\begin{cases} a = 1 \text{ for absorption} \\ a = 2 \text{ for reflection} \end{cases}$

But for photons: $p = E/c$, and: $F = \frac{a}{\Delta t}\frac{E}{c}$

and since: $\frac{E}{\Delta t}$ = power = P , finally: $\boxed{F = \frac{aP}{c}}$

Now: 50% of the radiation is absorbed: $F_1 = 0.5\,\frac{P}{c}$

50% is reflected: $F_2 = 0.5\,\frac{2P}{c}$

Hence: $F = F_1 + F_2 = 0.5\,\frac{P}{c}\,(1+2) = \frac{3}{2}\,\frac{P}{c}$

Numerically: $F = \frac{3}{2}\,\frac{8\ (\text{joule/sec})}{3 \times 10^8\ (\text{m/sec})} = 4 \times 10^{-8}\ \frac{\text{newton-m}}{\text{m}}$

or: $\boxed{F = 4 \times 10^{-8}\ \text{newton}}$ Answer

Now, let: N = photons/unit area x unit time
I = energy/unit area x unit time
$h\nu$ = energy/photon

$\therefore\ I = N h\nu$

But if A is the cross section of the beam, then: $I = P/A$, and:

$$\frac{P}{A} = N h\nu, \quad \text{or:} \quad N = \frac{P}{Ah\nu}$$

Since the photons move with speed c, the beam density is:

$$n = \frac{N}{c} = \frac{P}{Ah\nu c} = \frac{P\lambda}{Ahc^2} \quad \frac{\text{photons}}{\text{unit volume}}$$

Numerically:

$$n = \frac{8 \times 10^7 (\text{erg/sec}) \times 5 \times 10^{-5} (\text{cm})}{10\ (\text{cm}^2) \times 6.626 \times 10^{-27} (\text{erg-sec}) \times 9 \times 10^{20} (\text{cm/sec})^2}$$

or: $\boxed{n = 6.7 \times 10^7\ \text{photons/cm}^3}$ Answer

****

29 1.5 ev photoelectrons are ejected from a tungsten surface by ultraviolet light. If the threshold of tungsten for photoelectric effect is 2300 Å, what is the wavelength of the incident radiation ?

Answer: 1800 Å.

****

30 To stop the flow of photoelectrons produced by electromagnetic radiation incident on a certain metal, a negative potential of 300 volts is required. If the photoelectric threshold of the metal is 1500A, what is the frequency of the incident radiation ? Is this radiation visible ?

Answers: $\nu = 7.45 \times 10^{16}$ cycles/sec ; no.

****

3

THE WAVE NATURE OF MATTER

31

Calculate the wavelength of the wave associated with: (a) a rifle bullet weighing 2 gm and moving with a speed v = 500 m/sec; (b) a 1-Mev photon; (c) a 1-Mev electron; and (d) a 1-Mev proton.

****

On the De Broglie's hypothesis, the wavelength associated with a particle is given by the equation:

$$\boxed{\lambda = \frac{h}{p} = \frac{h}{mv}} \qquad (1)$$

where: h = Plank constant = 6.626×10^{-27} erg sec.
p = momentum of particle

(a) Here: $m = 2$ gm, $v = 5 \times 10^4$ cm/sec. Thus:

$$\lambda = \frac{6.626 \times 10^{-27}}{2 \times 5 \times 10^4} = \boxed{6.626 \times 10^{-32} \text{ cm}} \qquad \underline{\text{Ans.(a)}}$$

(b) For photons,

$$p = \frac{h\nu}{c} = \frac{E_\gamma}{c}$$

and equation (1) becomes:

$$\lambda = \frac{hc}{E_\gamma} = \frac{6.626 \times 10^{-27} \times 3 \times 10^{10}}{1 \times 1.6 \times 10^{-6}} = \boxed{1.242 \times 10^{-10} \text{ cm}} \qquad \underline{\text{Ans.(b)}}$$

(c) It is necessary in this case to calculate the momentum from relativistic formulae. Thus:

$$E^2 = p^2c^2 + (m_oc^2)^2 = (K + m_oc^2)^2 = K^2 + (m_oc^2)^2 + 2K\, m_oc^2$$

$$\therefore \quad p = \frac{1}{c}\sqrt{K(K + 2m_oc^2)} = \frac{1}{c}\sqrt{1(1 + 1.02) \times (1.6 \times 10^{-6})^2}$$

$$= \frac{1.6 \times 10^{-6}\sqrt{2.02}}{3 \times 10^{10}} = 7.58 \times 10^{-17} \text{ gm cm/sec}$$

and:

$$\lambda = \frac{6.626 \times 10^{-27}}{7.58 \times 10^{-17}} = \boxed{8.74 \times 10^{-11} \text{ cm}} \qquad \underline{\text{Ans.(c)}}$$

d) It is easy to see that a 1-Mev proton is not relativistic. From:

$$K = (\gamma - 1)E_o$$

one gets:

$$\gamma = \frac{K}{E_o} + 1 = \frac{1}{938.2} + 1 \approx 1$$

Then: $p = \sqrt{2mK}$, and (1) becomes:

$$\lambda = \frac{h}{\sqrt{2mK}} = \frac{6.626 \times 10^{-27}}{\sqrt{2 \times 1.6725 \times 10^{-24} \times 1 \times 1.6 \times 10^{-6}}}$$

or:

$$\boxed{\lambda = 2.87 \times 10^{-12} \text{ cm}} \qquad \underline{\text{Ans.(d)}}$$

****

32 The molecules of a certain gas at 320°K have a root-mean-square speed of 0.499 Km/sec. Calculate the De Broglie wavelength of these molecules. What is the gas ?

****

The average kinetic energy of the molecules is expressed:

$$\langle E \rangle = \frac{1}{2} mv^2 = \frac{3}{2} kT \qquad (1)$$

where: k = Boltzmann constant = 1.38×10^{-16} erg/°K.

Thus:

$$p = mv = \frac{3kT}{v}$$

and since the De Broglie wavelength is: $\lambda = \frac{h}{p}$

one has:

$$\lambda = \frac{hv}{3kT}$$

Using CGS units:

$$\lambda = \frac{6.626 \times 10^{-27} (\text{erg-sec}) \times 0.499 \times 10^{5} \ (\text{cm/sec})}{3 \times 1.38 \times 10^{-16} (\text{erg/°K}) \times 320 \ (\text{°K})} = 2.5 \times 10^{-9} \text{ c}$$

or: $\boxed{\lambda = 0.25 \ \text{Å}}$ <u>Answer</u>

It is quite easy to see that the gas is oxygen. From (1):

$$m = \frac{3kT}{v^2} = \frac{3 \times 1.38 \times 10^{-16} \times 320}{(0.499)^2 \times 10^{10}} (\text{gm}) \times \frac{1}{1.66 \times 10^{-24}} \left(\frac{\text{amu}}{\text{gm}}\right)$$

$$m = 32 \text{ amu}$$

and the gas is O^{16}.

****

33 A beam of neutrons with energies ranging from zero to several electron volts is directed at a crystal with a grating space of 3.03 Å (calcite).

(a) Determine the angle between the incident beam and the crystal so that the reflected neutrons will have a kinetic energy of 0.1 ev.

(b) Calculate the wavelength, in Angstroms, of the following:
1. the neutrons reflected by the crystal
2. electrons having the same energy as the reflected neutrons
3. electrons having the same momentum as the reflected neutrons.

****

(a) Bragg's law is written:

$$n\lambda = 2d \sin \theta \qquad (1)$$

where λ, the De Broglie wavelength, is given by:

$$\lambda = \frac{h}{p} = \frac{h}{\sqrt{2mK}} \qquad (2)$$

By substitution of (2) into (1):

$$\frac{nh}{\sqrt{2mK}} = 2d \sin \theta$$

and from here:

$$\sin \theta = \frac{nh}{2d\sqrt{2mK}}$$

Numerically, using $m = 1.675 \times 10^{-27}$ Kg for the mass of the neutron:

$$\sin \theta = \frac{1 \times 6.626 \times 10^{-34}}{2 \times 3.03 \times 10^{-10}\sqrt{2 \times 1.675 \times 10^{-27} \times 0.1 \times 1.6 \times 10^{-19}}}$$

or: $\sin \theta = 0.1495$

$$\boxed{\theta = 8°36'}$$ Ans.(a)

(b) Using equation (1):

$$\lambda = 2 \times 3.03 \times 0.1495$$

or: $$\boxed{\lambda = 0.906 \text{ Å}}$$ Ans.(b)1.

Since electrons with energy K = 0.1 ev are not relativistic, the corresponding wavelength can be obtained using equation (2); using MKS units:

$$\lambda = \frac{h}{\sqrt{2mK}} = \frac{6.626 \times 10^{-34}}{\sqrt{2 \times 9.1 \times 10^{-31} \times 0.1 \times 1.6 \times 10^{-19}}} (m) \times 10^{+10} \ (\text{Å}/m)$$

$$\boxed{\lambda = 38.9 \text{ Å}}$$ Ans.(b)2.

Finally, if the electron has the same momentum than the neutron, it will also have the same wavelength, and:

$$\boxed{\lambda = 0.905 \text{ Å}}$$ Ans.(b)3.

****

34

(a) Determine the wavelength of an electron which has been accelerated through a potential of 511 Kev.
(b) What would be the wavelength of a photon having the same energy as the electron ?
(c) What would be the energy, in Mev, of a photon having the same wavelength as the electron ?

****

(a) Since the rest energy of the electron is $E_o = 511$ Kev, this electron is clearly relativistic and has a total energy:

$$E = E_o + K = 0.511 + 0.511 = 2 \times 0.511 \text{ Mev}$$

Hence:

$$\gamma = \frac{E}{E_o} = 2$$

and:

$$\beta = \frac{\sqrt{\gamma^2 - 1}}{\gamma} = 0.866$$

The De Broglie wavelength of the electron is then:

$$\lambda = \frac{h}{p} = \frac{h}{m\,v} = \frac{h}{\gamma m_o \beta c}$$

Using CGS units:

$$\lambda = \frac{6.626 \times 10^{-27} \text{ (erg-sec)}}{2 \times 9.1 \times 10^{-28} \text{(gm)} \times 3 \times 10^{10} \text{ (cm/sec)}} = 1.4 \times 10^{-10} \text{ cm}$$

or: $\boxed{\lambda = 1.4 \times 10^{-2} \text{ Å}}$ Ans.(a)

(b) We have now a photon of energy: $E_\gamma = 0.511$ Mev

But: $E_\gamma = pc$, thus: $p = \frac{E_\gamma}{c}$

and its wavelength is: $\lambda = \frac{h}{p} = \frac{hc}{E_\gamma}$ (1)

Numerically:

$$\lambda = \frac{6.626 \times 10^{-27} \text{(erg-sec)} \times 3 \times 10^{10} \text{ (cm/sec)}}{0.511 \text{ (Mev)} \times 1.6 \times 10^{-6} \text{ (erg/Mev)}} = 2.438 \times 10^{-10} \text{ cm}$$

or: $\boxed{\lambda = 2.438 \times 10^{-2} \text{ Å}}$ Ans.(b)

(c) Using equation (1) above:

$$E_\gamma = \frac{hc}{\lambda} = \frac{6.626 \times 10^{-27} \times 3 \times 10^{10}}{1.4 \times 10^{-10}} \text{ (erg)} \times \frac{1 \text{ (Mev)}}{1.6 \times 10^{-6} \text{(erg)}}$$

or: $\boxed{E_\gamma = 0.886 \text{ Mev}}$ Ans.(c)

****

35 The electrons of a beam incident on a crystal at an angle of 30° have kinetic energies ranging from zero to a maximum of 5500 ev. The crystal has a grating space d = 0.5 A, and the reflected electrons are passed through a slit as shown in the figure. Find the velocities of the electrons passsing through the slit. How many are these velocities ?

****

Using the De Broglie relation:

$$\lambda = \frac{h}{p} = \frac{h}{mv}$$

one obtains:

$$v = \frac{h}{m\lambda} \qquad (1)$$

Now, Bragg's law is: $n\lambda = 2d \sin \theta = 2d \sin 30° = 2d \times \frac{1}{2} = d$

or: $\lambda = \frac{d}{n}$

and by substitution into (1):

$$\boxed{v = \frac{nh}{md}} \qquad (2)$$

The electrons passing through the slit will have therefore several velocities, corresponding to the values n= 1, 2,...provided those velocities do not exceed the maximum velocity in the beam. Let's calculate that maximum velocity. The most energetic electrons have:

$$K = 5500 \text{ ev} = 5.5 \text{ Kev}$$

and thus: $E = E_o + K = 511 + 5.5 = 516.5$ Kev

Hence: $\gamma = \frac{E}{E_o} = 1 + \frac{K}{E_o} = 1 + \frac{5.5}{511} = 1 + 0.01078 = 1.01078 \simeq 1.011$

Notice that by proceeding in this manner the slide rule error in the calculation of γ is minimized. Compare with the value that is obtained by dividing directly, $\gamma = 516.5/511$.

For this value of γ, one has: $\beta = 0.148$, and the maximum velocity present in the beam is:

$$v = \beta c = 0.148 \times 3 \times 10^{10} = 4.44 \times 10^{9} \text{ cm/sec}$$

Returning to equation (2), if the electrons are relativistic, the mass should be replaced by the relativistic mass, $m = \gamma m_o$, which depends on the velocity. Consequently, it is better to use (2) to calculate momentum, and from there obtain the velocity.

$\boxed{n = 1}$ From (2): $p = mv = \frac{nh}{d}$, and: $pc = n \frac{hc}{d}$

Using CGS units:

$$pc = \frac{6.626 \times 10^{-27} \times 3 \times 10^{10}}{0.5 \times 10^{-8}} \text{(erg)} \times \frac{1 \text{ (Mev)}}{1.6 \times 10^{-6} \text{ (erg)}}$$

or: $pc = 2.48 \times 10^{-2}$ Mev

Then, the total energy is: $E^2 = p^2c^2 + E_o^2 = (2.48)^2 \times 10^{-4} + (0.511)^2$

$$\text{or:} \quad E^2 = 6.15 \times 10^{-4} + (0.511)^2 \approx (0.511)^2 = E_o^2$$

Hence, the relativistic effects can be ignored in this case. Using (2), then:

$$v = \frac{1 \times 6.626 \times 10^{-27}}{9.1 \times 10^{-28} \times 0.5 \times 10^{-8}} = \boxed{1.457 \times 10^9 \quad \text{cm/sec}}$$ **Answer**

$\boxed{n = 2}$ Similarly:

$$pc = 2 \times 2.48 \times 10^{-2} = 4.96 \times 10^{-2} \text{ Mev}$$

Thus: $E^2 = p^2c^2 + E_o^2 = (4.96)^2 \times 10^{-4} + (0.511)^2 = 0.00246 + 0.26112$

$$\text{or:} \quad E^2 = 0.26358 \quad (\text{Mev})^2$$

But: $\beta^2 = 1 - \dfrac{E_o^2}{E^2} = \dfrac{E^2 - E_o^2}{E^2} = \dfrac{p^2c^2}{E^2} = \dfrac{0.00246}{0.26358} = 0.00993$

$$\text{or:} \quad \beta = 0.0996$$

and the velocity of the electrons is:

$$v = \beta c = 0.0996 \times 3 \times 10^{-10} = \boxed{2.988 \times 10^9 \quad \text{cm/sec}}$$ **Answer**

Notice that if formula (2) is used directly:

$$v = 2 \times 1.457 \times 10^9 = 2.914 \times 10^9 \quad \text{cm/sec}$$

which is practically the same result, indicating that also for this case the relativistic effects are small. This is to be expected, since $v < 0.1c$.

$\boxed{n = 3}$ Similarly:

$$pc = 3 \times 2.48 \times 10^{-2} = 7.44 \times 10^{-2} \text{ Mev}$$

and: $E^2 = p^2c^2 + E_o^2 = (7.44)^2 \times 10^{-4} + (0.511)^2 = 0.005535 + 0.26112$

$$\text{or:} \quad E^2 = 0.26665 \ (\text{Mev})^2$$

$$\therefore \quad \beta^2 = \frac{p^2c^2}{E^2} = \frac{0.005535}{0.26665} = 0.02075 \qquad \text{and:} \qquad \beta = 0.144$$

The velocity is then:

$$v = \beta c = 0.144 \times 3 \times 10^{10} = \boxed{4.32 \times 10^9 \ \text{cm/sec}}$$ **Answer**

$\boxed{n = 4}$ Similarly:

$$pc = 4 \times 2.48 \times 10^{-2} = 9.92 \times 10^{-2} \text{ Mev}$$

$$E^2 = p^2c^2 + E_o^2 = 0.00984 + 0.26112 = 0.27096 \ (\text{Mev})^2$$

$$\beta^2 = \frac{p^2c^2}{E^2} = \frac{0.00984}{0.27096} = 0.0363, \quad \text{or: } \beta = 0.190$$

which exceeds the maximum value of β previously calculated. Consequently, only electrons of three different velocities will pass through the slit., corresponding to $n = 1, 2,$ and 3.

****

36

The continuous X-ray spectrum from a molybdenum target is analyzed with a calcite crystal spectrometer (grating space of calcite is 3.03Å). The smallest angle at which an intense spot is observed on the screen is 10°.

(a) What is the shortest wavelength of the spectrum ?

(b) If copper had been used instead of molybdenum, how would the shortest wavelength be affected ?

(c) What is the accelerating potential in the X-ray tube ?

****

(a) Using Bragg's law:

$$\boxed{n\lambda = 2d \sin \theta}$$

where: $n = 1$
$d = 3.03$ Å
$\theta = 10°$, or $\sin \theta = \sin 10° = 0.1736$

one obtains:

$$\lambda = 2 \times 3.03 \times 0.1736 \text{ Å}$$

or:

$$\boxed{\lambda = 1.05 \text{ Å}}$$

Ans.(a)

(b) The shortest wavelength corresponds to the maximum energy of the X-rays, which in turn depends on the energy of the electrons incident on the target, but not on the substance of the target. Hence, the shortest wavelength would be the same.

(c) The maximum X-ray energy corresponds to the maximum kinetic energy of the electrons. If V is the accelerating potential, then:

$$Ve = h\nu = \frac{hc}{\lambda}$$

and from here:

$$V = \frac{hc}{\lambda e}$$

and using MKS units:

$$V = \frac{6.626 \times 10^{-34}(\text{joule-sec}) \times 3 \times 10^{8}(\text{m/sec})}{1.05\ (\text{Å}) \times 10^{-10}(\text{m/Å}) \times 1.6 \times 10^{-19}(\text{coul})}$$

or:

$$\boxed{V = 11.85 \text{ Kv}}$$

Ans.(c)

****

4

PARTICLES IN FIELDS

37

Electrons emitted with negligible energy from a hot filament are accelerated toward an electrode which is 1,000,000 volts positive with respect to the filament. (a) What is the kinetic energy of the electrons after being accelerated ? (b) What is the final velocity of the electrons ?

****

By the definition of electron volt, one can write at once the kinetic energy of electrons accelerated through a potential difference of 10^6 volts:

$$\boxed{K = 1 \text{ Mev}}$$ Ans.(a)

Of course, the same result can be obtained from first principles. The kinetic energy is equal to the work done by the electric force due to the field:

$$K = Fd = Ee\ d = \frac{V}{d} e\ d = Ve$$

where E is the electric field and d is the distance between the electrodes. Using CGS electrostatic units:

$$K = \frac{10^6}{300} \text{ (statvolts) } \times 4.8 \times 10^{-10} \text{ (esu) } = 1.6 \times 10^{-6} \text{ ergs}$$

and we have actually shown that:

$$\boxed{1 \text{ Mev} = 1.6 \times 10^{-6} \text{ ergs}}$$

(b) Using the relativistic formula:

$$K = m_o c^2(\gamma - 1)$$

one obtains:

$$\gamma = \frac{K}{m_o c^2} + 1 = \frac{1}{0.511} + 1 = 1.96 + 1 = 2.96$$

and using tables:

$$\beta = 0.941$$

Then:

$$v = \beta c = 0.941 \times 3 \times 10^{10} = \boxed{2.823 \times 10^{10} \text{ cm/sec}}$$ Ans.(b)

****

38 A 10 Mev electron is moving perpendicularly to a magnetic field of 5000 gauss. Determine the radius of curvature of its path.

****

The centripetal force is provided by the magnetic field, and the electron moves on a circular path of radius r:

$$F_m = Bev = m\frac{v^2}{r}$$

Now, in this formula, if B is expressed in gauss, the electronic charge should be expressed in the corresponding electromagnetic units (i.e., abcoulombs). Recalling that the ratio of the electrostatic and electromagnetic units of charge is 1/c, the above equation can be rewritten as follows:

$$\frac{Bev}{c} = \frac{mv^2}{r}$$

where now: $e = 4.8 \times 10^{-10}$ esu. Then:

$$r = \frac{mvc}{Be} = \frac{pc}{Be} \qquad (1)$$

The relativistic momentum is expressed in terms of γ and β :

$$p = \gamma m_o \beta c = \frac{\gamma\beta}{c}\, m_o c^2$$

and (1) finally becomes:

$$\boxed{r = \frac{\gamma\beta m_o c^2}{Be}} \qquad (2)$$

The value of γ is obtained from the equation:

$$K = m_o c^2(\gamma - 1)$$

Hence: $$\gamma = 1 + \frac{K}{m_o c^2} = 1 + \frac{10}{.511} = 1 + 19.6 = 20.6$$

and: $\beta = 0.998$

Replacing numerical values into (2):

$$r = \frac{20.6 \times 0.998 \times .511(\text{Mev}) \times 1.6 \times 10^{-6}(\text{erg/Mev})}{5 \times 10^{3}\ (\text{gauss}) \times 4.8 \times 10^{10}\ (\text{esu})}$$

or: $$\boxed{r = 7\ \text{cm}}$$ Answer

****

39

A beam of electrons is deflected through a semicircular path of radius r = 10 cm by means of a magnetic field of 5000 gauss. Calculate the kinetic energy of the electrons.

****

The centripetal force is provided by the magnetic field and:

$$Bev = \frac{mv^2}{r}, \qquad \text{or:} \qquad p = mv = Ber$$

Using CGS units, but expressing e in esu instead of emu, above formula is written:

$$p = \frac{Ber}{c}, \qquad \text{or:} \qquad pc = Ber \quad \text{ergs}$$

Numerically:

$$p = \frac{5000\ (\text{gauss}) \times 4.8 \times 10^{-10}\ (\text{esu}) \times 10\ (\text{cm})}{1.6 \times 10^{-6}\ (\text{erg/Mev})} = 15\ \text{Mev}$$

Now:

$$E^2 = p^2c^2 + E_o^2 = (15)^2 + (0.511)^2 = 225 + 0.261 = 225.26$$

$$\therefore \quad E = \sqrt{225.26} = 15.01\ \text{Mev}$$

and since: $E = K + E_o$, finally: $K = E - E_o = 15.01 - 0.511 = 14.5$ Mev

or: $\boxed{K = 14.5\ \text{Mev}}$ Answer

****

40

Non-relativistic particles of charge e, mass m, and velocity v are sent through the curved trajectory shown in the figure, where E is the electric field at right angles to the path.

(a) Show that this device acts as an energy filter, i.e., that the particles emerging at A all have the same kinetic energy.

(b) If the same set up is used except that a magnetic field B at right angles to the figure replaces the electric field, show that the device acts as a momentum filter.

(c) Discuss how (a) and (b) could be combined to determine e/m for the particles.

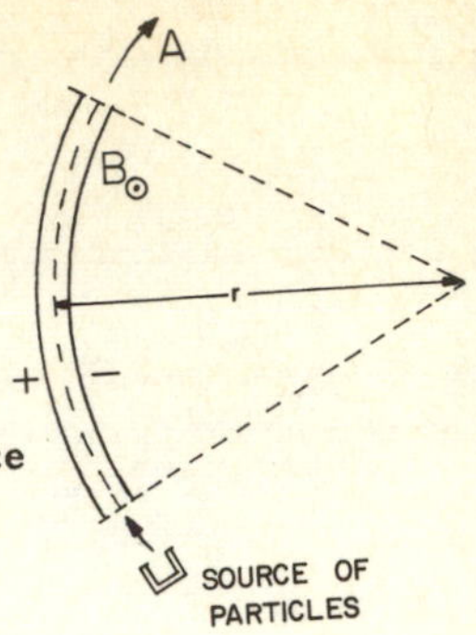

****

(a) It is assumed that the separation between the curved surfaces is small compared to the radius r of the path. If the particles are positively charged, and the electric field is established as indicated in the figure, the electric force will provide the centripetal acceleration required for the particles to follow a circular path:

$$eE = \frac{mv^2}{r} \tag{1}$$

If the particles were negatively charged, the same effect would be obtained by reversing the polarity of the plates.

Since E, e, and r are fixed, equation (1) indicates that only particles with the same mv^2, i.e., the same kinetic energy, will be able to traverse the device and emerge at A.

(b) When the electric field is substituted by a magnetic field perpendicular to the paper, the centripetal force will be provided by the magnetic force:

$$Bev = \frac{mv^2}{r}$$

or:

$$Ber = mv = p \tag{2}$$

Since B, e, and r are fixed, this equation indicates that only particles of momentum p will traverse the device. Again, if the particles are positively charged the field B should be directed out of the figure, as shown, while if the particles are negatively charged the same effect is obtained with a field directed into the paper.

(c) Consider now the case when particles of charge e, mass m, and velocity v (where $v \ll c$) are emitted by the source. If a magnetic field is established, the particles will not pass through unless the field B has the value that satisfies equation (2). By changing B until particles are observed emerging at A, one determines the velocity of the particles to be:

$$v = \frac{e}{m} Br \tag{3}$$

Next, the magnetic field is replaced by an electric field E, which again is adjusted until the particles emerge at A. When this is the case, equation (1) is satisfied and:

$$\frac{e}{m} = \frac{v^2}{rE} \tag{4}$$

By substitution of (3) in (4):

$$\frac{e}{m} = \frac{1}{rE}\frac{e^2}{m^2}B^2r^2$$

or:

$$\boxed{\frac{e}{m} = \frac{E}{B^2 r}}$$

where the values of E and B are obtained experimentally as discussed above.

****

41

In the arrangement shown below, the first crystal has a grating space of 1 Å, while the grating space of the second crystal is 0.8 Å.

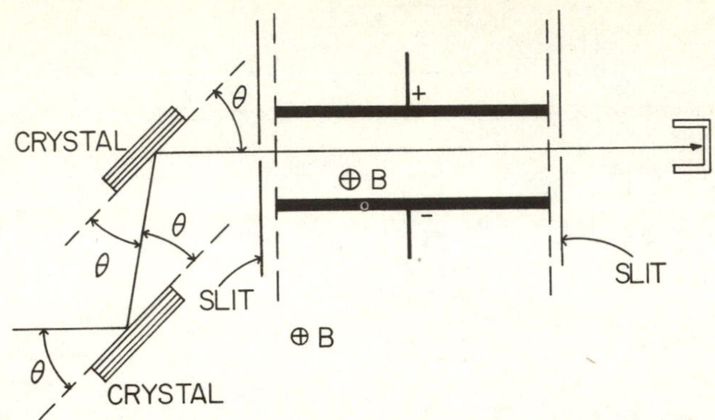

A beam of non-relativistic particles is incident upon the first crystal and reflected toward the second crystal. All angles θ are equal, and θ = 30°. After the second reflection, the beam enters a region of crossed electric and magnetic fields, where: E = 19.8 kilovolts/meter, and B = 1 weber/m^2, directed into the figure. The slits are such that allow the passage only of undeviated particles.

(a) The beam contains charged particles of several velocities and masses, with positive and negative charges. Specify the characteristics (mass, charge, velocity) of the particles that will reach the detector.

(b) The beam contains now only neutrons of energies ranging from zero to 1 ev. Could any of these neutrons reach the detector ?

****

(a) Bragg's law for a crystal is written:

$$n\lambda = 2d \sin \theta \qquad (1)$$

where d is the grating space, and λ, the De Broglie wavelength, is given by:

$$\lambda = \frac{h}{p} \qquad (2)$$

Combining (1) and (2), and considering that θ = 30°, or, sin θ = 1/2, one has:

$$\boxed{p = \frac{nh}{d}} \qquad (3)$$

Consequently, the first crystal will pass particles with momentum:

$$p_1 = n \frac{6.626 \times 10^{-34} (\text{joule-sec})}{1\ (\text{A}) \times 10^{-10}\ (\text{A/m})} = 6.626 \times 10^{-24}\ n \quad \frac{\text{joule-sec}}{\text{meter}}$$

or, since 1 joule = 1 newton x 1 meter,

$$p_1 = 6.626 \times 10^{-24}\ n \qquad \text{newton-sec}$$

Likewise, the second crystal will pass particles having momentum:

$$p_2 = 6.626 \times 10^{-24} \frac{n'}{0.8} \quad \text{newton-sec}$$

and the arrangement of the two crystal acts as a momentum filter, passing only particles with momentum:

$$p = p_1 = p_2, \qquad \text{or:} \qquad n = \frac{n'}{0.8}$$

Hence:

$$\frac{n'}{n} = 0.8 = \frac{4}{5}$$

and the lowest values of n and n' that will satisfy this condition are:

$$n = 5, \qquad \text{and:} \qquad n' = 4 .$$

The particles reaching the crossed fields will have momentum:

$$p = 6.626 \times 10^{-24} \times \frac{4}{0.8} = \boxed{3.313 \times 10^{-23} \text{ newton-sec}}$$

Now, in the region of the crossed fields, the electric and magnetic forces are directed in opposite directions, and the particles will be allowed to pass only when these two forces are equal:

$$\left.\begin{aligned} F_e &= eE \\ F_m &= eBv \end{aligned}\right\} \quad \text{or:} \qquad v = \frac{E}{B} \qquad (4)$$

The device acts as a velocity filter. Notice that if the sign of the particles is changed, both forces F_e and F_m will change direction, so that the filter is not dependent on the charge.
Replacing numerical values into (4):

$$v = \frac{19800}{1} \frac{(\text{volts})}{(\text{wb/m}^2)} = \boxed{1.98 \times 10^4 \text{ m/sec}}$$

The mass of the particles is obtained as follows:

$$p = mv \qquad \therefore \qquad m = \frac{p}{v} = \frac{3.313 \times 10^{-23}}{1.98 \times 10^4} \left(\frac{\text{newton-sec}}{\text{m/sec}}\right)$$

$$\text{or:} \boxed{m = 1.672 \times 10^{-27} \text{ Kg}}$$

Therefore: particles reaching detector are protons (or anti-protons) with velocity $v = 1.98 \times 10^4$ m/sec.

(b) The neutrons are not affected by the velocity filter, but the momentum filter will operate as before. The minimum momentum required to pass the filter is, as previously calculated:

$$p = 3.313 \times 10^{-23} \text{newton-sec}$$

and the corresponding minimum kinetic energy is:

$$E = \frac{p^2}{2m} = \frac{(3.313)^2 \times 10^{-46}}{2 \cdot \times 1.675 \times 10^{-27}} (\text{joules}) \times \frac{1 \text{ (ev)}}{1.6 \times 10^{-19} \text{ (joules)}} = 2.04 \text{ ev}$$

Since the maximum energy of the neutrons in the beam is 1 ev, no neutron will reach the detector.

****

42

A beam of positively charged particles enters the device shown below, where the magnetic field is uniform and directed out of the paper. The electric plates are parallel and very close, so that the electric field is given by V/d, and directed perpendicular to the path of the particles. Find an expression for the distance x from the origin at which a particle returns to the x-axis in terms of particle charge e, mass m, velocity v, B, d, and V.

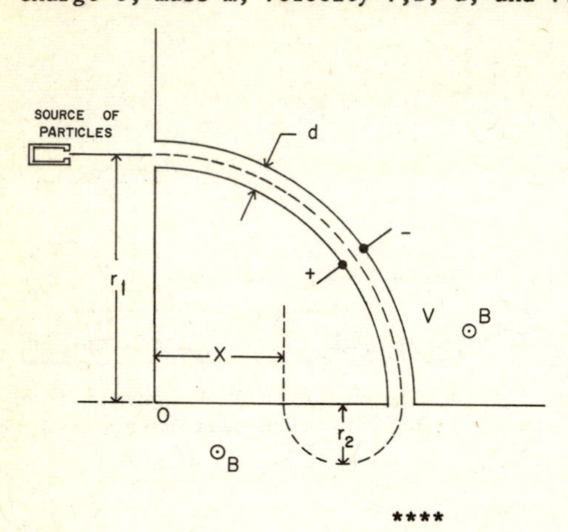

****

As the particles enter the device, they are acted upon by a magnetic force F_m, directed toward the center O since the particles are positively charged, and an electric force, F_e, directed outward. Thus, the centripetal force is:

$$\frac{mv^2}{r_1} = F_m - F_e = Bev - eE$$

$$\therefore \quad r_1 = \frac{mv^2}{Bev - eE} \qquad (1)$$

In the lower part, after leaving the electric field:

$$\frac{mv^2}{r_2} = Bev, \qquad \therefore \quad r_2 = \frac{mv^2}{Bev} \qquad (2)$$

From the geometry:

$$x = r_1 - 2r_2$$

and by substitution of (1) and (2):

$$x = \frac{mv^2}{e(Bv-E)} - \frac{2mv^2}{Bev}$$

or:

$$x = \frac{mv^2}{e}\left[\frac{Bv - 2(Bv-E)}{(Bv-E)Bv}\right] = \frac{mv}{Be}\left[\frac{2E - Bv}{Bv - E}\right]$$

and since $E = V/d$, finally:

$$\boxed{x = \frac{mv}{Be}\left(\frac{2V - Bvd}{Bvd - V}\right)}$$

****

43

A beam of 1 Mev electrons enters a region of uniform crossed electric and magnetic fields. The magnetic field is directed into the paper, and extends over the region indicated in the figure. The strength of the electric field is $E = 1.335 \times 10^6$ volts/meter.

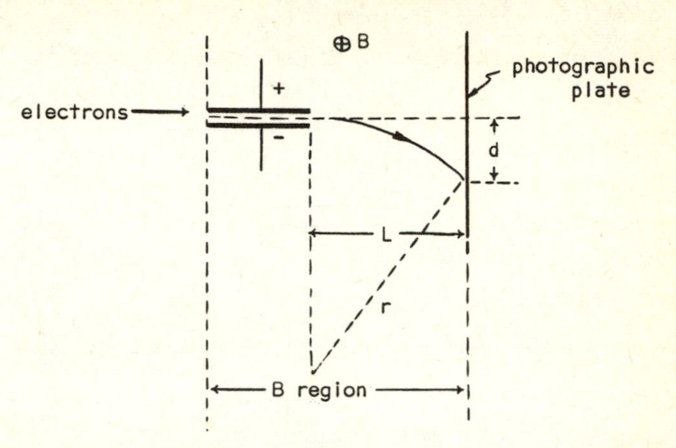

(a) Show that the radius of the circular path followed by the electrons after leaving the parallel plates is given by the expression:

$$r = \frac{m_o c^2}{eE} \frac{\beta^2}{\sqrt{1 - \beta^2}}$$

where $\beta = v/c$, and m_o = electron rest mass.

(b) Calculate the value of r. If L = 10 cm, at what distance d will the electrons strike the photographic plate ?

****

(a) The cross fields act as a velocity selector. Equating the electric and magnetic forces:

$$eE = Bev , \qquad \text{or:} \qquad v = \frac{E}{B} \qquad (1)$$

After leaving the parallel plates, the centripetal acceleration of the electrons is provided by the magnetic force and:

$$\frac{mv^2}{r} = Bev, \qquad \text{or:} \qquad r = \frac{mv}{Be} \qquad (2)$$

where m is the relativistic mass: $m = \gamma m_o$, and B can be obtained from (1). Equation (2) becomes then:

$$r = \frac{\gamma m_o v}{eE/v} = \frac{\gamma m_o v^2}{eE}$$

and finally:

$$\boxed{r = \frac{m_o c^2}{eE} \frac{\beta^2}{\sqrt{1 - \beta^2}}} \qquad (3) \qquad \text{Q.E.D.}$$

(b) The value of γ is obtained by using the relation:

$$K = m_o c^2(\gamma - 1) \quad \therefore \quad \gamma = 1 + \frac{K}{m_o c^2} = 1 + \frac{1}{0.511} = 2.96$$

Then: $\beta = 0.94$, and $\beta^2 = 0.884$.

By substitution of numerical values in (3), using MKS units:

$$r = \frac{2.96 \times 0.511 \text{ (Mev)} \times 1.6 \times 10^{-13} \text{(joule/Mev)}}{1.6 \times 10^{-19} \text{ (coul)} \times 1.335 \times 10^{6} \text{ (volt/m)}}$$

or: $\boxed{r = 1 \text{ meter}}$ Answer

Now, from the geometry: $r^2 = L^2 + (r-d)^2$

or: $d = r - \sqrt{r^2 - L^2} = 1 - \sqrt{1 - 0.01} = 1 - \sqrt{0.99} = 1 - 0.995 = 0.005$

or: $\boxed{d = 5 \text{ mm}}$ Answer

****

44 In the arrangement shown below, the region to the left and above a'b'c' has a uniform magnetic field B, directed out of the paper. Outside this region, the magnetic field is zero. The electric field between the two parallel plates is E. Field fringing and relativity effects are negligible.

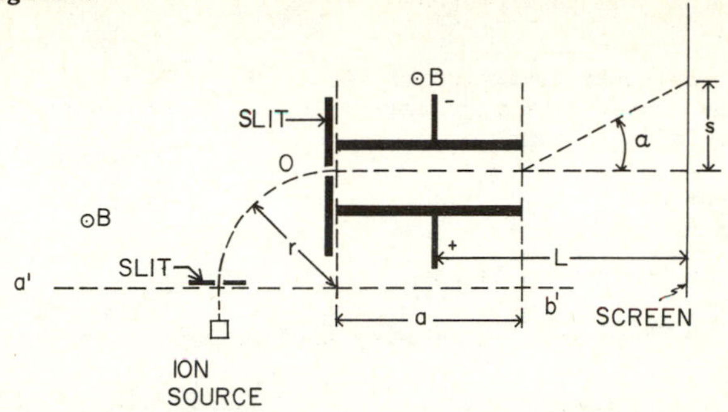

The ion source produces singly charged ions having different masses and velocities. Derive an equation giving the deflection s at which the ions of different masses will strike the screen in terms of the various parameters of the problem. Assume that the two slits are very narrow, and that the deflection angle α is small.

****

In the region where only the magnetic field B is present, one has:

$$Bev = \frac{mv^2}{r} , \qquad \text{or:} \qquad mv = Ber$$

indicating that this part of the device acts as a momentum filter. The ions will arive to the point 0 with horizontal velocity:

$$v = v_x = \frac{Ber}{m} \qquad (1)$$

In the region where both electric and magnetic fields are present the resultant force acting upon the ions is:

$$F = eE - Bev = ma_y$$

and the vertical acceleration a_y is therefore:

$$a_y = \frac{dv_y}{dt} = \frac{e(E-Bv)}{m}$$

By integration, the vertical velocity is expressed:

$$v_y = \frac{e(E-Bv)}{m} t \qquad (2)$$

which, when the particles leave the region between the parallel plates, i.e., at:

$$t = \text{transit time} = \frac{a}{v_x} = \frac{a}{v}$$

takes the value: $$v_y = \frac{e(E-Bv)a}{mv} \tag{3}$$

Integrating (2) again:

$$y = \frac{e(E-Bv)t^2}{2m}$$

which at $t = a/v$ gives the deviation of the particles from the center line when they emerge from the parallel plates:

$$AB = \frac{e(E-Bv)a^2}{2mv^2}$$

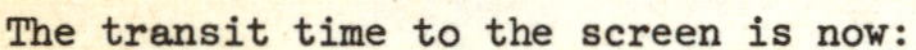

The transit time to the screen is now:

$$t_1 = \frac{L - (a/2)}{v} \quad \text{(assuming the angle } \alpha \text{ is small)}$$

and the corresponding vertical displacement is, using (3):

$$BC = v_y t_1 = \frac{e(E-Bv)a}{mv} \left[\frac{L - (a/2)}{v} \right]$$

The total vertical deviation is then:

$$s = AB + BC = \frac{e(E-Bv)}{2mv^2}[a^2 + 2a(L- \frac{a}{2})] = \frac{e(E-Bv)aL}{mv^2}$$

Finally, replacing v by its expression (1):

$$\boxed{s = \frac{aL(mE - B^2er)}{B^2er^2}}$$

****

45

Antiprotons with velocity 0.68c enter a magnetic spectrometer operating with a field B = 1.45 weber/m^2, and are turned through an angle of 90°.
(a) What is the momentum of the antiprotons in Mev/c ?
(b) Calculate the radius of the path of the particles.
(c) If the momentum of the antiprotons varies by 0.5%, what will be the spatial separation observed as the particles leave the spectrometer ?

****

(a) The antiprotons have: $\beta = 0.68$

Then (using tables): $\gamma = 1.365$

and their momentum is:

$$p = mv = \gamma m_o \beta c = \frac{\gamma\beta}{c} m_o c^2$$

Since the rest energy of the antiprotons is: $m_o c^2 = 938.25$ Mev, one has:

$$p = \frac{1.365 \times 0.68 \times 938.25(\text{Mev})}{c}$$

or: $\boxed{p = 0.87 \times 10^3 \text{ Mev/c}}$ Ans.(a)

(b) The centripetal acceleration is provided by the magnetic force:

$$\frac{mv^2}{r} = Bev \text{ , and from here: } \quad r = \frac{mv}{Be}$$

Using CGS EM units, but expressing the charge of the electron in esu, this formula becomes:

$$r = \frac{pc}{Be} = \frac{0.87 \times 10^3(\text{Mev}) \times 1.6 \times 10^{-6}(\text{erg/Mev})}{1.45 \times 10^4(\text{gauss}) \times 4.8 \times 10^{-10}(\text{esu})}$$

or: $\boxed{r = 200 \text{ cm}}$ Ans.(b)

(c) Since: $r = \frac{p}{Be}$

then: $\Delta r = \frac{\Delta p}{Be}$

and dividing these two relations: $\frac{\Delta r}{r} = \frac{\Delta p}{p}$ $\therefore$ $\Delta r = r \frac{\Delta p}{p}$

Here: $\frac{\Delta p}{p} = 0.005$, $r = 200$ cm

Then: $\Delta r = 2 \times 10^2 \times 5 \times 10^{-3} = 1$ cm

or $\boxed{\Delta r = 1 \text{ cm}}$ Ans.(c)

Notice that this separation could be doubled by turning the particles through an angle of 180°.

****

46

A beam of electrons having speeds ranging from zero to 0.5c passes between two parallel plates separated by a distance d = 0.02 cm and 10 cm long. The upper plate has a potential of +30 volts relative to the lower plate, and to stop all electrons except those of a specified velocity, a magnetic field is applied perpendicular to the direction of the beam.

(a) What magnetic field is required to pass electrons having a speed of 0.01c ?

(b) Because the distance between the plates is not infinitesimal, electrons with a small range of speeds will be passed. Calculate approximately $\Delta v/v$ when $v = 0.01c$.

(c) Assume now that the magnetic field is 10 gauss. What are the mass and the kinetic energy of the electrons that will pass through the device?

****

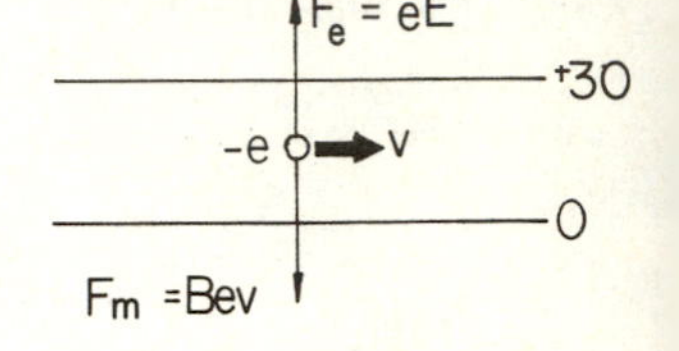

(a) The electrons will be allowed to pass the device when the electric force, directed upwards, is balanced by the magnetic force. Thus, the magnetic field is directed into the paper and:

$$eE = Bev \quad , \quad \text{or: } B = \frac{E}{v}$$

Using MKS units:

$$B = \frac{30 \text{ (volts)}}{2 \times 10^{-4} \text{ (m)} \times 0.01 \times 3 \times 10^{8} \text{ (m/sec)}} = \frac{1}{20} \frac{\text{weber}}{\text{m}^2}$$

or, since $1 \text{ wb/m}^2 = 10^4$ gauss:

$$\boxed{B = 500 \text{ gauss}}$$

Ans.(a)

(b) If the velocity of the electrons is $(v+\Delta v)$ the net force acting on them will be:

$$F = -eE + Bev + Be\,\Delta v = Be\,\Delta v$$

which depending on the sign of Δv will be directed upward or downward. This force will produce a vertical acceleration:

$$a = \frac{F}{m} = \frac{Be}{m}\,\Delta v$$

which will act during the transit time of the electron through the device:

$$t = \frac{L}{(v+\Delta v)} \simeq \frac{L}{v}$$

and will produce a deflection:

$$y = \frac{1}{2}at^2 = \frac{1}{2}\frac{Be}{m}\,\Delta v\,\frac{L^2}{v^2}$$

For an electron to pass, this deflection can be at most equal to the plate separation d, and:

$$d = \frac{Be}{2m}\,\frac{L^2}{v^2}\,\Delta v$$

and from here:

$$\boxed{\frac{\Delta v}{v} = \frac{2d}{Be}\,\frac{mv}{L^2}}$$

Numerically, for $v = 0.01c$:

$$\frac{\Delta v}{v} = \frac{2 \times 2 \times 10^{-2}(\text{cm}) \times 9.1 \times 10^{-28}(\text{gm}) \times 10^{-2} \times 3 \times 10^{10}(\text{cm/sec})}{500\ (\text{gauss}) \times 4.8 \times 10^{-10}\ (\text{esu}) \times (1/3 \times 10^{10})(\text{esu/abcoul}) \times 10^{2}(\text{cm}^2)}$$

or: $$\boxed{\frac{\Delta v}{v} = 1.36 \times 10^{-5}}$$ Ans.(b)

(c) The electric field is still the same:

$$E = \frac{V}{d} = \frac{30}{2 \times 10^{-4}} = 1.5 \times 10^{5} \text{ volts/meter}$$

but the magnetic field is now:

$$B = 10 \text{ gauss} = 10^{-3} \text{ weber/m}^2$$

Thus, the velocity of the electrons passing the device is:

$$v = \frac{E}{B} = \frac{1.5 \times 10^{5}}{10^{-3}} = 1.5 \times 10^{8} \text{ meter/sec}$$

or: $$v = \frac{c}{2}$$

Using tables, for $\beta = 0.500$ one gets: $\gamma = 1.155$, and:

$m = \gamma m_0 = 1.155 \times 9.1 \times 10^{-28}$ gm, or: $\boxed{m = 1.05 \times 10^{-27} \text{ gm}}$

$K = E_0(\gamma - 1) = 0.511 \times 0.155$ Mev, or: $\boxed{K = 0.0792 \text{ Mev}}$

Answers (c)

****

5

ATOMIC PHYSICS

47

The particles in a narrow beam are scattered by a thin foil of thickness t, having n atoms per unit volume. Consider that the scattering is the result of collisions between perfectly elastic spheres, and that there are no interparticle fields such as those of electrical or nuclear forces. The radius of the bombarding particles is R_1 and the radius of the target atoms is R_2.
Derive expressions for:

(a) the differential cross section for scattering into an angular interval $d\phi$ about ϕ.

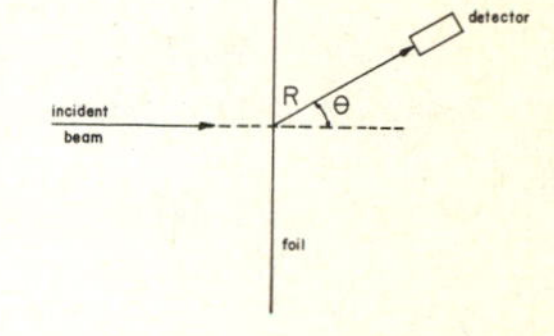

(b) the cross section $\sigma(\phi)$ for scattering through an angle larger than any specified angle ϕ.

(c) the probability that a particle will be scattered by the foil through an angle larger than ϕ.

(d) the probability, per unit area of detector, that a single particle will be scattered into a detector located at a distance R and at an angle ϕ.

****

(a) The foil is assumed to be thin enough so that there is no shadowing between the scattering centers in different layers. The incident particles will either pass through the foil undeviated, or they will be scattered if they pass within a distance b, the impact parameter, from the scattering center. Since the collision is elastic, the angle of incidence and the angle of reflection, θ, relative to the line joining the centers are equal, and the angle of deflection, ϕ, is:

$$\phi = \pi - 2\theta$$

or: $$\theta = \frac{\pi}{2} - \frac{\phi}{2}$$

The impact parameter is then:

$$b = (R_1+R_2)\sin\theta = (R_1+R_2)\cos\frac{\phi}{2}$$

All particles approaching the scattering center with impact parameter in db about b will be deflected into an angle $d\phi$ about ϕ, or, in other words, all particles incident upon the differential cross section:

$$d\sigma = 2\pi b\, db$$

will be scattered into the differential solid angle:

$$d\Omega = 2\pi \sin\phi\, d\phi$$

The differential cross section per unit solid angle is then:

$$\frac{d\sigma}{d\Omega} = \frac{2\pi b\, db}{2\pi \sin\phi d\phi} \qquad (2)$$

From (1):

$$db = -(R_1+R_2)\frac{1}{2}\sin\frac{\phi}{2}\, d\phi$$

Here, the - sign simply indicates that if b increases, ϕ decreases, and will be ignored. Equation (2) becomes then:

$$\frac{d\sigma}{d\Omega} = \frac{(R_1+R_2)\cos\frac{\phi}{2}\,(R_1+R_2)\,\frac{1}{2}\sin\frac{\phi}{2}\,d\phi}{\sin\phi\,d\phi}$$

or:

$$\frac{d\sigma}{d\Omega} = \frac{1}{4}(R_1+R_2)^2 \qquad \text{(say, cm}^2\text{/steradian)}$$

Thus, the differential cross section per unit solid angle is constant, indicating that the scattering is isotropic.

Now:

$$d\sigma = \frac{1}{4}(R_1+R_2)^2\,d\Omega = \frac{1}{4}(R_1+R_2)^2 2\pi\sin\phi\,d\phi$$

Or:

$$\boxed{d\sigma = \frac{\pi}{2}(R_1+R_2)^2\sin\phi\,d\phi} \qquad (3) \qquad \underline{\text{Ans.(a)}}$$

(b) The cross section for scattering through an angle larger than ϕ is obtained by integration of (3) from ϕ to π:

$$\sigma(\phi) = \int_\phi^\pi d\sigma = \frac{\pi}{2}(R_1+R_2)^2\int_\phi^\pi \sin\phi\,d\phi = \frac{\pi}{2}(R_1+R_2)^2\Big[-\cos\phi\Big]_\phi^\pi$$

or:

$$\boxed{\sigma(\phi) = \pi(R_1+R_2)^2\cos^2\frac{\phi}{2}} \qquad (4) \qquad \underline{\text{Ans.(b)}}$$

Notice that in terms of b, this integration is between $b = 0$ (i.e, $\phi = \pi$) and $b = b(\phi)$,(i.e., $\phi = \phi$). In other words, the limits of integration have been reversed. This is because of the - sign that was dropped before.

(c) If the foil has area A, thickness t, and contains n scattering centers per unit volume, the probability for a single particle to be scattered into $d\phi$ about ϕ is:

$$df = \frac{nAt\,d\sigma}{A} = nt\,d\sigma = \frac{nt\pi}{2}(R_1+R_2)^2\sin\phi\,d\phi \qquad (5)$$

This expressions gives, in other words, the fraction of the incident particles scattered into $d\phi$ about ϕ. The fraction that is scattered through an angle larger than ϕ is then:

$$\boxed{f = nt\sigma(\phi) = nt\pi\,(R_1+R_2)^2\cos^2\frac{\phi}{2}}$$

(d) The area intersected by a sphere of radius R between the two cones of angles ϕ and $\phi + d\phi$ is;

$$a = 2\pi r\,ds = 2\pi R\sin\phi\,R d\phi$$

or: $$a = 2\pi R^2\sin\phi\,d\phi \qquad (6)$$

Hence, the probability that a single particle will be scattered into the detector, per unit area of detector, is obtained by dividing expressions (5) and (6):

$$\boxed{p = \frac{nt}{4}\left(\frac{R_1+R_2}{R}\right)^2} \qquad \underline{\text{Ans.(d)}}$$

****

48

An alpha particle that has been accelerated through a potential difference of 100,000 volts is incident normally on a foil of copper 0.001 millimeter thick. Copper has a density of 8.8 gm/cm^3. Calculate the probability that the alpha particle will be scattered (i.e., deflected) through an angle less than 60°.

****

The classical theory of Rutherford scattering† shows that the probability for an alpha particle to be deflected through an angle larger than θ is:

$$f = nt\sigma$$

where: n = number of scattering centers per unit volume
t = thickness of the foil
σ = cross section = πb^2

The impact parameter, b, depends on the angle of scattering and on the atomic number of the scatterer:

$$b = \frac{2Ze^2}{mv^2} \cot\frac{\theta}{2}$$

Combining all these expressions, the probability for deflection through an angle larger than θ is expressed:

$$\boxed{f = \pi nt\left(\frac{2Ze^2}{mv^2}\right)^2 \cot^2\frac{\theta}{2}} \qquad (1) \qquad \text{(CGS units)}$$

In the present case:

$$n = \frac{N_o \rho}{W} \qquad \text{where:} \begin{cases} N_o = \text{Avogadro's number} \\ W = \text{atomic weight of copper} \end{cases}$$

Thus: $$n = \frac{6.023 \times 10^{23} \times 8.8}{63.54} = 8.34 \times 10^{22} \quad \text{atoms/cm}^3$$

Also: $$\theta = 60° \therefore \cot\frac{\theta}{2} = \cot 30° = \sqrt{3} \quad ; \quad Z = 29 \text{ (for copper)}$$

$$\frac{mv^2}{2} = 0.2 \text{ Mev} = 0.2 \times 1.6 \times 10^{-6} \text{ ergs}$$

Replacing numerical values in (1), then:

$$f = \pi \times 8.34 \times 10^{22} \times 10^{-4} \left[\frac{29 \times (4.8)^2 \times 10^{-20}}{0.2 \times 1.6 \times 10^{-6}}\right]^2 \times 3 = 0.0343$$

The probability for deflection through an angle less than 60° is therefore:

$$p = 1 - f = 1 - 0.0343 = 0.9657$$

or: $$\boxed{p = 96.57\%}$$ Answer

****

(†) See for example, Leighton, Principles of Modern Physics, McGraw-Hill, 1959, page 485.

49

(a) Determine the cross section in barns (1 barn = 10^{-24} cm^2) for the scattering of 1 Mev alpha particles through an angle larger than 90° in a thin foil of copper.

(b) What fraction of the alpha particles will be scattered through an angle larger than 90°, if the foil is 0.01 mm thick ?

For copper: Z = 29, W = 63.54, and ρ = 8.9 gm/cm^3.

****

(a) The cross section is: $\sigma = \pi b^2$

where b, the impact parameter, is given by the expression:

$$b = \frac{Ze^2}{E_k} \cot \frac{\theta}{2}$$

Here: $\theta = 90° \quad \therefore \quad \frac{\theta}{2} = 45° \quad$ and: $\cot \frac{\theta}{2} = \cot 45° = 1$

E_k = 1 Mev = 1.6 x 10^{-6} ergs

Then:

$$\sigma = \pi \left[\frac{29 \times (4.8)^2 \times 10^{-20} \times 1}{1.6 \times 10^{-6}}\right]^2 = \pi \times (4.18)^2 \times 10^{-24} \text{ cm}^2$$

or: $\boxed{\sigma = 55 \text{ barns}}$ Ans(a)

(b) The fraction of the alpha particles that will be scattered through and angle larger than θ is:

$$f = nt\sigma(\theta)$$

where: n = number of scattering centers per unit volume = $\frac{N_o \rho}{W}$
t = thickness of foil

Then:

$$f = \frac{6.023 \times 10^{23} \times 8.9 \times 10^{-3} \times 55 \times 10^{-24}}{63.54} = 0.00464$$

or: $\boxed{f = 0.46 \%}$ Ans.(b)

*****

50 A 0.5 Mev alpha particle approaches the nucleus of an iron atom (Z = 26) which is at rest. The impact parameter is $b = 10^{-10}$ cm.
(a) Find the angle through which the alpha particle is scattered.
(b) Determine the distance of closest approach to the nucleus.

****

(a) The impact parameter b and the angle of scattering θ are related by the expression (in CGS units):

$$\boxed{b = \frac{2Ze^2}{mv^2}\cot\frac{\theta}{2}}$$

Then: $$\tan\frac{\theta}{2} = \frac{2Ze^2}{mv^2 b} = \frac{26 \times (4.8)^2 \times 10^{-20}}{0.5 \times 1.6 \times 10^{-6} \times 10^{-10}} = 0.0749$$

or: $\frac{\theta}{2} = 4°17'$ ∴ $\boxed{\theta = 8°34'}$ Ans.(a)

(b) The potential energy is:

$$V = -\int_{\infty}^{r} F_c dr = -\int_{\infty}^{r} \frac{2Ze^2}{r^2} dr = -\left[-\frac{2Ze^2}{r}\right]_{\infty}^{r}$$

or: $$V(r) = \frac{2Ze^2}{r}$$

which is zero for $r = \infty$. Thus, at infinity, the total energy of the particle is equal to its kinetic energy:

$$E = K_{\infty} = 0.5 \text{ Mev}$$

At the point of closest approach, the potential energy is:

$$V(r_o) = \frac{2Ze^2}{r_o}$$

Hence, by conservation of energy:

$$E = K(r_o) + \frac{2Ze^2}{r_o}$$

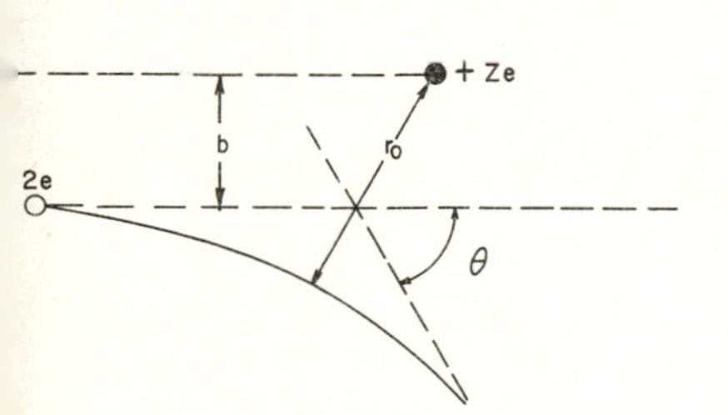

or: $$K_o = K(r_o) = E - \frac{2Ze^2}{r_o} \qquad (1)$$

Now, the angular momentum when the particle is at infinity is:

$$L_{\infty} = mvb = \sqrt{2mK_{\infty}}\, b = \sqrt{2mE}\, b$$

while at closest approach:

$$L_o = \sqrt{2mK_o}\, r_o$$

$$\sqrt{2mE}\, b = \sqrt{2mK_o}\, r_o$$

and using (1): $$Eb^2 = K_o r_o^2 = \left[E - \frac{2Ze^2}{r_o}\right] r_o^2$$

or: $$\frac{Eb^2}{r_o^2} + \frac{2Ze^2}{r_o} - E = 0$$

Solving for $1/r_o$: $$\frac{1}{r_o} = \frac{-2Ze^2 \pm \sqrt{4Z^2e^4 + 4E^2b^2}}{2Eb^2}$$

Since $r_o > 0$, only the + sign has physical significance and:

$$\frac{1}{r_o} = -\frac{Ze^2}{Eb^2} + \sqrt{\left(\frac{Ze^2}{Eb^2}\right)^2 + \frac{1}{b^2}}$$

or:

$$\frac{1}{r_o} = \frac{Ze^2}{Eb^2}\left[-1 + \sqrt{1 + \left(\frac{Eb}{Ze^2}\right)^2}\,\right]$$

Numerically: $$\frac{Eb}{Ze^2} = \frac{0.5 \times 1.6 \times 10^{-6} \times 10^{-10}}{26 \times (4.8)^2 \times 10^{-20}} = 13.35$$

$$\left(\frac{Eb}{Ze^2}\right)^2 = (13.35)^2 = 178.3$$

Then: $$\frac{1}{r_o} = \frac{26 \times (4.8)^2 \times 10^{-20}}{0.5 \times 1.6 \times 10^{-6} \times 10^{-20}}\left[-1 + \sqrt{1 + 178.3}\right]$$

or: $$r_o = \frac{0.5 \times 1.6 \times 10^{-6}}{26 \times (4.8)^2 \times 12.4}$$

and finally: $$\boxed{r_o = 1.08 \times 10^{-10} \text{ cm}}$$ Ans.(b)

****

51

A Bohr hydrogen atom at rest in free space undergoes a transition from the energy level $n = 2$ to $n = 1$, emitting a photon in the positive x direction. Since this photon has momentum, the atom must recoil in the negative x direction. Using conservation of energy and momentum, find the recoil velocity of the atom. (Hint: you may need to use the first two terms of the series expansion: $(1+x)^{1/2} = 1 + x/2 - x^2/8 + \ldots$)

****

The energy levels for the hydrogen atom are given, in CGS units, by:

$$\boxed{E_n = -\frac{me^4}{2n^2\hbar^2}}$$

and for $n = 1$, $E_1 = -13.58$ ev. Thus, for $n = 2$:

$$E_2 = \frac{E_1}{n^2} = -\frac{13.58}{4} = -3.39 \text{ ev}$$

and:

$$\Delta E = E_2 - E_1 = -3.39 + 13.58 = 10.19 \text{ ev}$$

By conservation of energy:

$$\frac{Mv^2}{2} + E_\gamma = \Delta E \qquad (1)$$

where: M = mass of hydrogen atom
E_γ = energy of emitted photon

By conservation of momentum:

$$Mv = p_\gamma = \frac{E_\gamma}{c}$$

or:

$$E_\gamma = Mvc$$

and replacing in (1):

$$\frac{Mv^2}{2} + Mvc = \Delta E$$

or:

$$v^2 + 2cv - \frac{2\Delta E}{M} = 0$$

Solving for v:

$$v = -c \pm \sqrt{c^2 + \frac{2\Delta E}{M}}$$

Since $|v| < |c|$, only the + sign has physical significance, and using the series expansion for the square root, as hinted, one has:

$$v = -c + c\sqrt{1 + \frac{2\Delta E}{Mc^2}} \simeq c[-1 + 1 + \frac{2\,\Delta E}{2Mc^2}] = \frac{\Delta E}{Mc}$$

Numerically:

$$M = 1.0078 \text{ amu} = 1.0078 \times 1.66 \times 10^{-24} \text{ gm}$$

$$\therefore \quad v = \frac{10.19(\text{ev}) \times 1.6 \times 10^{-12}(\text{erg/ev})}{1.0078 \times 1.66 \times 10^{-24}(\text{gm}) \times 3 \times 10^{10}(\text{cm/sec})}$$

or:

$$\boxed{v = 325 \text{ cm/sec}}$$

Answer

****

52

(a) On the basis of the Bohr model, calculate the third ionization potential of lithium.
(b) The electron in a doubly ionized lithium atom drops from n = 2 to the ground state. Find the energy, the momentum and the wavelength of the emitted photon. Is this radiation within the visible range ?

****

(a) Doubly ionized lithium, Li^{++}, has a single electron bound to a Z = 3 nucleus, i.e., it is a hydrogenic atom. The energy levels are obtained from those of hydrogen by replacing e^2 by Ze^2:

$$\boxed{E_n = -\frac{mZ^2e^4}{2n^2\hbar^2}} \qquad (1)$$

The third ionization potential is the energy required to remove the remaining electron from its ground state, i.e., when n = 1 in (1):

$$E_I = -E_1 = \frac{mZ^2e^4}{2\hbar^2}$$

Since: $\frac{me^4}{2\hbar^2}$ = ionization potential of hydrogen = 13.58 ev

it follows that: $E_I = Z^2 \times 13.58 = 9 \times 13.58$

or: $\boxed{E_I = 122 \text{ ev}}$ Answer (a)

(b) The energy levels (1) can be better expressed:

$$E_n = \frac{E_1}{n^2} \qquad \text{where:} \quad E_1 = -\frac{mZ^2e^4}{2\hbar^2} = -122 \text{ ev}$$

Then, the transition from n = 2 to n = 1 will result in the emission of a photon of energy:

$$E_\gamma = h\nu = E_2 - E_1 = \frac{E_1}{4} - E_1 = -\frac{3}{4}E_1 = \frac{3}{4} \times 122$$

or: $\boxed{E_\gamma = 91.5 \text{ ev}}$ Answer

The corresponding wavelength is:

$$\lambda = \frac{hc}{E_\gamma}$$

which, if E_γ is expressed in ev and the wavelength is in Å, is also written:

$$\lambda = \frac{1.24 \times 10^4}{E_\gamma(\text{ev})} \text{ Å} = \frac{1.24 \times 10^4}{91.5} \text{ Å}$$

or: $\boxed{\lambda = 135.5 \text{ Å}}$ not visible. Answer

Finally, the momentum of the photon is:

$$p_\gamma = \frac{E_\gamma}{c} = \frac{91.5(\text{ev}) \times 1.6 \times 10^{-12}(\text{erg/ev})}{3 \times 10^{10} \text{ (cm/sec)}}$$

or: $\boxed{p_\gamma = 4.88 \times 10^{-21} \text{ gm cm/sec}}$ Answer

****

53

(a) From considerations of equilibrium, derive the following relation between the orbital radius and the angular velocity of the electron in doubly ionized lithium:

$$r = \frac{912}{\omega^{2/3}} \qquad \text{(CGS units)}$$

(b) Using the formula of part (a), apply Bohr's postulate on angular momentum to find the minimum orbital radius.

****

(a) Let: m = mass of electron
M = mass of Li nucleus

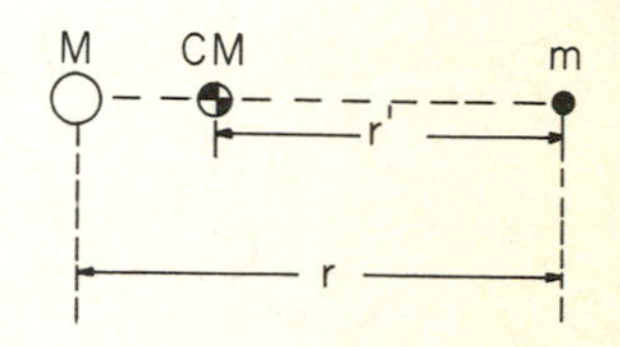

The particles revolve about the center of mass, which is at a distance r' from the electron, such that:

$$(M+m)r' = Mr \quad , \qquad \text{or: } r' = \frac{M}{M+m}\, r$$

Equating the centripetal force to the Coulomb attractive force:

$$\frac{mv^2}{r'} = \frac{Ze^2}{r^2}$$

or, since $v = \omega r'$:

$$\frac{Ze^2}{r^2} = m\omega^2 r' = \omega^2 r \frac{Mm}{M+m}$$

Define:

$$\mu = \frac{Mm}{M+m} = \frac{m}{1+\frac{m}{M}}$$

as the reduced mass of the system, and observe that since M>>m, numerically $\mu \simeq m$. Then:

$$\frac{Ze^2}{r^2} = \mu\,\omega^2 r$$

or:

$$r = \frac{1}{\omega^{2/3}} \sqrt[3]{\frac{Ze^2}{\mu}}$$

Here:

$$\sqrt[3]{\frac{Ze^2}{\mu}} = \sqrt[3]{\frac{3 \times (4.8)^2 \times 10^{-20}}{9.1 \times 10^{-28}}} = \sqrt[3]{7.6 \times 10^8} = 912$$

and:

$$\boxed{r = \frac{912}{\omega^{2/3}}} \qquad (1) \qquad\qquad \underline{\text{Q.E.D.}}$$

(b) In terms of the reduced mass, the total angular momentum of the system is $\mu r^2\omega$, and using Bohr Second postulate:

$$\mu r^2 \omega = n\hbar$$

or, since $\mu \simeq m$, $\qquad m^2 r^4 \omega^2 = n^2\hbar^2$

Now, from (1): $\qquad \omega^2 = \frac{(912)^3}{r^3}$

and substituting above and solving for r:

$$r = \frac{n^2\hbar^2}{(912)^3 m^2}$$

Numerically:

$$r = \frac{(6.626)^2 \times 10^{-54}}{7.6 \times 10^8 \times 4\pi^2 \times (9.1)^2 \times 10^{-56}} = 0.177 \times 10^{-8} \text{ cm}$$

or: $\boxed{r = 0.176 \text{ Å}}$ <u>Answer</u>

****

54 Consider a hypothetical one-electron atom which does not have the energy levels of hydrogen, but which is assumed to obey Bohr's third postulate. The wavelengths of the first four spectral lines of the series terminating on n = 1 are: 1200Å, 1000Å, 900Å, and 840Å. The shortest wavelength of the series is 800Å.

(a) Determine the first five energy levels of this atom in ev, and draw an energy level diagram. What is the ionization potential ?
(b) Calculate the wavelength of the line corresponding to the transition from n = 3 to n =2 .
(c) What is the minimum energy to be supplied to the electron in the ground state so that the transition in part (b) will be possible ?

****

(a) The atom satisfies Bohr's Third postulate, i.e.,

$$\boxed{h\nu = E_i - E_f}$$

where E_i is the energy of the initial state and E_f the energy of the final state. The shortest wavelength will correspond to the transition from $n = \infty$, i.e, $E_i = 0$, to the ground state, n = 1:

$$E_1 = -h\nu = -\frac{hc}{\lambda}$$

To simplify the computations, one can write:

$$\frac{hc}{\lambda} = \frac{6.626 \times 10^{-27}\,(\text{erg-sec}) \times 3 \times 10^{10}(\text{cm/sec})}{\lambda(\text{A}) \times 10^{-8}\,(\text{cm/A}) \times 1.6 \times 10^{-12}(\text{erg/ev})} = \frac{1.24 \times 10^4}{\lambda(\text{A})} \quad \text{ev}$$

which gives the energies directly in ev when the wavelengths are expressed in Å. Thus, for $\lambda = 800$ Å:

$$E_1 = -\frac{1.24 \times 10^4}{800} = -15.5 \text{ ev}$$

For the other transitions terminating on n = 1, one has:

$$h\nu = E_n - E_1$$

or:

$$E_n = E_1 + \frac{hc}{\lambda} = -15.5 + \frac{1.24 \times 10^4}{\lambda(\text{Å})} \quad \text{ev}$$

Hence:

$$E_2 = -15.5 + \frac{1.24 \times 10^4}{1200} = -15.5 + 10.33 = -5.17 \text{ ev}$$

$$E_3 = -15.5 + \frac{1.24 \times 10^4}{1000} = -15.5 + 12.40 = -3.10 \text{ ev}$$

$$E_4 = -15.5 + \frac{1.24 \times 10^4}{900} = -15.5 + 13.77 = -1.73 \text{ ev}$$

$$E_5 = -15.5 + \frac{1.24 \times 10^4}{840} = -15.5 + 14.77 = -0.73 \text{ ev}$$

The ionization potential is: $\boxed{E_I = -E_1 = 15.5 \text{ ev}}$ Answer

(b) For the transition from n = 3 to n = 2:

$$h\nu = E_3 - E_2 = -3.10 + 5.17 = 2.07 \text{ ev}$$

Then: $\lambda = \dfrac{hc}{h\nu}$

$$= \frac{6.626 \times 10^{-27} \times 3 \times 10^{10}}{2.07 \times 1.6 \times 10^{-12}} \text{ cm}$$

or: $\boxed{\lambda = 6000 \text{ Å}}$ Answer

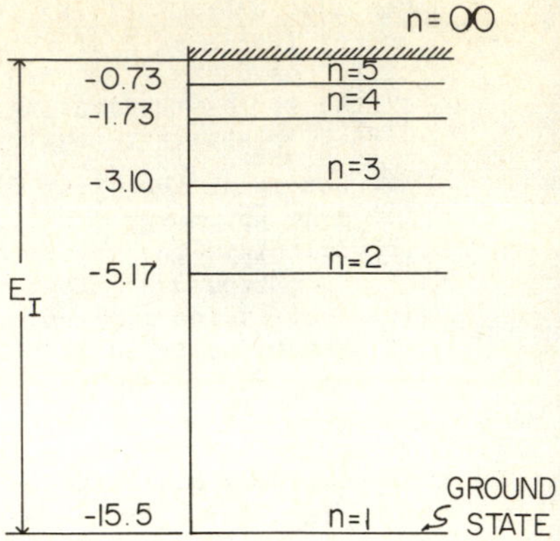

n = ∞
-0.73 n=5
-1.73 n=4
-3.10 n=3
-5.17 n=2
E_I
-15.5 n=1 GROUND STATE

(c) For the above transition to be possible, the atom must be excited to the n = 3 state. Thus, the minimum energy is:

$$\Delta E = E_3 - E_1 = -3.10 + 15.5 \text{ ev}$$

or: $\boxed{\Delta E = 12.40 \text{ ev}}$ Answer

****

55

A μ meson is a particle with all the properties of the electron but a mass 207 times larger. Find its energy levels and orbit radii in the Coulomb field of a nucleus of charge Ze. For what value of Z is the muon inside the nucleus ? Assume that $Z = A/2$, where A is the mass number, and that the nuclear radius is given by $r = r_0 A^{1/3}$, where $r_o = 1.4 \times 10^{-13}$ cm. In particular, if the muon is bound to a proton, compute the energy of the ground state and the radius of the first orbit.

****

A μ^- meson has the same charge as the electron, but a mass 207 times larger. After losing energy by ionization and excitation processes, a muon can be captured by the Coulomb field of a nucleus into Bohr-type orbits, forming a *mesonic atom*. As will be seen below, the muon orbits are well inside the K-shell, and the particles will spend part of their time inside the nucleus itself. The influence of the orbital electrons is negligible, and the mesonic atomic behaves as a one-electron atom. According to the Bohr model, then, the energy levels are (in CGS units):

$$E_n = - \frac{m_r Z^2 e^4}{2n^2 \hbar^2} \qquad (1)$$

while the radii of the orbits are given by:

$$r_n = \frac{n^2 \hbar^2}{m_r Z e^2} \qquad (2)$$

where: m_r = reduced mass = $\dfrac{m_\mu}{1 + \dfrac{m_\mu}{M}}$

m_μ = mass of muon = 207 m

m = mass of electron

M = mass of nucleus

But: $\dfrac{m_\mu}{M} \simeq \dfrac{207\ m}{A\ m_p} = \dfrac{1}{A} \dfrac{207}{1836} << 1$, as soon as A becomes large enough, and the reduced mass becomes: $m_r = 207\ m$

Equations (1) and (2) are then written:

$$\boxed{E_n = - 207 \frac{m Z^2 e^4}{2n^2 \hbar^2}} \qquad (3)$$

and:

$$\boxed{r_n = \frac{1}{207} \frac{n^2 \hbar^2}{m Z e^2}} \qquad (4)$$

indicating that the energy levels are 207 times deeper than those of the hydrogenic atom, while the radii of the orbits are 207 times smaller.

The first orbital radius is obtained from (4) for n = 1:

$$r_1 = \frac{1}{207 Z} \frac{\hbar^2}{m e^2}$$

The quantity $\hbar^2/me^2$ is the radius of the first hydrogen orbit, and is usually represented in the literature by the symbol a_o:

$$a_o = \frac{\hbar^2}{me^2} = 0.529 \times 10^{-8} \text{ cm}$$

Thus:

$$r_1 = \frac{a_o}{207Z}$$

Now, the nuclear radius is: $r = r_o A^{1/3} = r_o (2Z)^{1/3}$

and equating these two equations:

$$r_o (2Z)^{1/3} = \frac{a_o}{207Z}$$

Solving for Z:
$$Z^4 = \frac{1}{2}\left(\frac{a_o}{207r_o}\right)^3 = \frac{1}{2}\left(\frac{0.529 \times 10^{-8}}{207 \times 1.4 \times 10^{-13}}\right)^3 = 3.064 \times 10^6$$

or:
$$\boxed{Z \simeq 42}$$ Answer

Hence, for $Z \simeq 42$ (molybdenum), the muon will be moving inside the nucleus.

In particular, if the muon is moving about a proton, its ground state energy is given by (3) for $Z = 1$, $n = 1$:

$$E_1 = - 207 \frac{me^4}{2\hbar^2} = -207 \times 13.58 \text{ ev}$$

or:
$$\boxed{E_1 = 2.82 \text{ Kev}}$$ Answer

The radius of the first orbit is obtained from (4):

$$r_1 = \frac{1}{207} \frac{\hbar^2}{me^2} = \frac{a_o}{207} = \frac{0.529 \times 10^{-8}}{207} \text{ cm}$$

or:
$$\boxed{r_1 = 2.55 \times 10^{-11} \text{ cm}}$$ Answer

****

56 Under certain circumstances, an electron and a positron may form a bound system known as positronium, in which the particles revolve about their center of mass.

(a) On the basis of the Bohr model, find an expression for the energy levels and the orbit radii of positronium.

(b) What is the energy of the ground state ?

(c) If the system undergoes a transition from n = 2 to n =1, is this radiation in the visible region ?

****

(a) Let r be the separation between the particles. The Coulomb force:

$$F_c = -\frac{e^2}{r^2} \qquad (1)$$

provides the centripetal force required so that the particles will move about the center of mass with velocity v:

$$\frac{mv^2}{r'} = \frac{e^2}{r^2}$$

But: $r' = r/2$, and:

$$mv^2 = \frac{e^2}{2r} \qquad (2)$$

The kinetic energy of each particle is $mv^2/2$, so this is also the total kinetic energy of the system:

$$K = \frac{e^2}{2r}$$

The potential energy is obtained from (1):

$$V = -\int F_c \, dr = e^2 \int_\infty^r \frac{dr}{r^2} = -\frac{e^2}{r}$$

The total energy of the system is then:

$$E = V + K = -\frac{e^2}{r} + \frac{e^2}{2r} = -\frac{e^2}{2r} \qquad (3)$$

Using now Bohr's Second postulate, we quantize the total angular momentum, which is:

$$2 \times mv\,r' = 2\,mv\,\frac{r}{2} = mvr$$

Thus:

$$n\hbar = mvr \qquad (4)$$

Eliminating the velocity between equations (3) and (4):

$$v^2 = \frac{e^2}{2mr} = \frac{n^2\hbar^2}{m^2r^2}$$

and solving for r:

$$\boxed{r_n = \frac{2n^2\hbar^2}{me^2}} \qquad (5)$$

which gives the quantized radii. Replacing into (3):

$$E_n = -\frac{e^2}{2}\,\frac{me^2}{2n^2\hbar^2}$$

or: $$\boxed{E_n = -\frac{me^4}{4n^2\hbar^2}} \qquad (6)$$

Notice that equations (5)and (6) can be obtained from the corresponding expressions for the hydrogen atom if the mass of the electron is replaced by the reduced mass, which in this case is:

$$m_r = \frac{m}{1 + \frac{m}{m}} = \frac{m}{2}$$

(b) For n = 1, equation (6) becomes:

$$E_1 = -\frac{1}{2}\,\frac{me^4}{2\hbar^2} = -\frac{1}{2} \times 13.58 \quad \text{ev}$$

since $me^4/2\hbar^2$ is recognized to be the ground state energy for hydrogen. Thus:

$$\boxed{E_1 = 6.76 \text{ ev}}$$ <u>Answer (b)</u>

(c) The energy of the photon emitted in the transition from n = 2 to n = 1 is:

$$h\nu = E_2 - E_1 = \frac{E_1}{4} - E_1 = -\frac{3}{4}E_1 = +\frac{3}{4} \times 6.76 = 5.07 \text{ ev}$$

The corresponding wavelength is then:

$$\lambda = \frac{1.24 \times 10^4}{E\ (\text{ev})} = \frac{1.24 \times 10^4}{5.07} \quad \text{Å}$$

or: $$\boxed{\lambda = 2450 \text{ Å}}$$ <u>Answer (c)</u>

which lies in the ultraviolet region.

****

57 An atom has an electron configuration given by: $1s^2 2s^2 2p^6 3s^1$.
(a) Identify the element.
(b) Select from the following list those configurations which in combination with the configuration given above represent permitted transitions. Explain.

i) $1s^2 2s^2 2p^6 3p^1$
ii) $1s^2 2s^2 2p^5 3s^2$
iii) $1s^2 2s^2 2p^6 3d^1$
iv) $1s^2 2s^2 2p^6 4p^1$
v) $1s^2 2s^1 2p^6 3s^2$
vi) $1s^2 2s^2 2p^6 4s^1$
vii) $1s^2 2s^2 2p^6 6p^1$
viii) $1s^1 2s^2 2p^6 3s^1 4s^1$

****

(a) The number of electrons in the atom is Z = 11. Hence, the element is sodium, and the atom is in the ground state.

(b) The permitted transitions must obey the rule: $\boxed{\Delta\ell = \pm 1}$ and terminate in the ground state.

Thus, (i), (ii), (iv) and (vii) represent permitted transitions, while the rest are forbidden for the reasons indicated below:

(iii) since: $\Delta\ell = 2$
(v), (vi), (viii) since: $\Delta\ell = 0$.

****

58

An atom has an electron configuration given by: $1s^2 2s^1$.
(a) Identify the element.
(b) Calculate the energy necessary to ionize this atom by removing the 2s (or valence) electron, assuming the atom to be hydrogen-like. However, the experimental value is 5.39 ev. Can you explain the difference with your result ?
(c) What is the effective charge, Z_{eff}, seen by the valence electron ?
(d) Calculate the approximate wavelength of the radiation resulting from the transition $3p \to 2s$. Is this radiation visible to the human eye ? What is the energy of the photon ?
(Hint: the ionization potential of hydrogen is 13.6 ev).

****

(a) The atom has three electrons, i.e., $Z = 3$, and the element is Li.

(b) For a hydrogen-like atom the last electron revolves about a core formed by the other electrons , which more or less shield the positive charges of the nucleus. If Z_{eff} represents the net charges acting on the valence electron,the energy levels can be obtained in terms of the energy levels of hydrogen:

$$E_n = Z^2_{eff} E^H_n = - Z^2_{eff} \frac{E^H_I}{n^2} \qquad (1)$$

In the present case, assuming that the 2s electron does not penetrate the core,

$$Z_{eff} = 3-2 = 1$$

while for hydrogen: $E^H_1 = - E^H_I = -13.6$ ev

and n = 2. Thus: $E^{Li}_I = E^H_I/4 = 13.6/4$ or: $\boxed{E^{Li}_I = 3.4 \text{ ev}}$ Ans.(b)

This value differs from the experimental value of 5.39 ev because the shielding is, of course, only approximate, and not complete as assumed above. The valence electron will see a Z_{eff} different from 1, as calculated below.

(c) From (1):

$$E_n = - Z^2_{eff} \frac{E^H_I}{n^2} = - E^{Li}_I$$

for n = 2. Then:

$$Z^2_{eff} = \frac{-n^2 E^{Li}_I}{-E^H_I} = \frac{4 \times 5.39}{13.6} = 1.587$$

or: $\boxed{Z_{eff} = 1.26}$ Ans.(c)

(d) For the transition $3p \to 2s$:

$$\Delta E = Z^2_{eff} E^H_I \left(\frac{1}{2^2} - \frac{1}{3^2}\right) = 1.587 \times 13.6 \times \frac{9-4}{4 \times 9}$$

$\boxed{\Delta E = h\nu = 2.99 \text{ ev}}$ Ans.(d)

The wavelength of the radiation is obatined by using the expression:

$$\lambda = \frac{1.24 \times 10^4}{E\ (ev)} \text{Å} = \frac{1.24 \times 10^4}{2.99} = \boxed{4150\ \text{Å}}$$ Ans.(d)

which is in the visible region.

****

59

Write out the complete electronic configuration for thallium (Z = 81). Express the lowest energy state using the conventional spectroscopic notation, giving also all the possible values of J.

****

The order in which the energy levels are filled is given by the well known scheme:

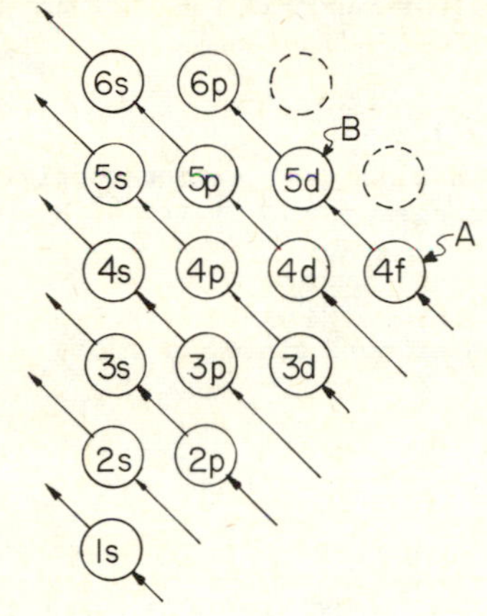

At point A, all the subshells for n = 4 are complete. The total number of electrons is:

for n = 1:	2	
" n = 2:	8	i.e., 60 electrons.
" n = 3:	18	
" n = 4:	32	

But some subshells for higher n are complete at point B:

subshell 5d:	10	
" 5p:	6	i.e., 20 electrons.
" 5s:	2	
and also subshell 6s:	2	

At point B, thus, one has 80 electrons. Since Tl has 81, the next electron is a $6p^1$, and the complete electronic configuration is written:

$$1s^2 2s^2 2p^6 3s^2 3p^6 4s^2 3d^{10} 4p^6 5s^2 4d^{10} 5p^6 6s^2 4f^{14} 5d^{10} 6p^1$$ Answer

The outer-most electron has $\ell = 1$ and since all other subshells are complete, this is also the orbital angular momentum of the atom: L = 1. Now, for the single electron, s = 1/2, so that for the atom the spin angular momentum is either S = 1/2, or S = -1/2. The total angular momentum:

$$J = L + S$$

takes 2S + 1 = 2 values, namely: $J = 1 + \frac{1}{2} = \frac{3}{2}$

and: $J = 1 - \frac{1}{2} = \frac{1}{2}$

The lowest energy state is expressed then: $6^2P_{1/2}$ Answer

****

60

Consider an excited atom of Cs (Z = 55). (a) Which one of the following transitions is possible:
(i) 7S → 6S, (ii) 7D → 6P, (iii) 6S → 5P ?
(b) Calculate the approximate wavelength of the possible transition in part (a).

****

(a) Cesium, with atomic number Z = 55, follows Xenon in the periodic table, and its electron configuration is:

$$1s^2 2s^2 2p^6 3s^2 3p^6 4s^2 3d^{10} 4p^6 5s^2 4d^{10} 5p^6 6s^1$$

i.e., it is formed by one 6s electron and a Xenon 54-electron configuration core. Thus, in the lower energy state, the extra electron is in the 6S state and the transition:

6S → 5P is not possible.

The transition: 7S → 6S is not possible either, since the rule:

$$\Delta \ell = \pm 1$$

is not satisfied. Hence, the only possible transition is:

$$\boxed{7D \rightarrow 6P}$$ Ans.(a)

(b) The atom can be considered as hydrogen-like and the approximate wavelength of the radiation calculated by using the expression:

$$\frac{1}{\lambda} = Z^2 R \left(\frac{1}{n_f^2} - \frac{1}{n_i^2} \right)$$

where:
$$R = R_\infty = 1.097 \times 10^{-3} \text{ Å}^{-1}$$
$$Z = Z_{eff} = 55 - 54 = 1$$
$$n_f = 6, \quad n_i = 7.$$

Thus:

$$\frac{1}{\lambda} = R \left(\frac{1}{36} - \frac{1}{49} \right) = \frac{13\,R}{36 \times 49}$$

or:
$$\lambda = \frac{36 \times 49}{1.097 \times 10^{-3} \times 13} \text{ Å}$$

and finally:
$$\boxed{\lambda = 1.24 \times 10^5 \text{ Å}}$$ Ans.(b)

which is in the infrared region.

****

61

Calculate the energy of the K_α X-rays of uranium, the element with highest Z found in nature. However, photons with energies larger than this value are observed. What is their origin ?

****

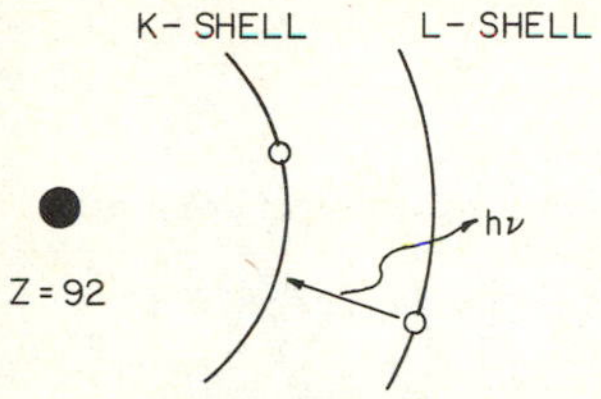

K- SHELL L- SHELL

$h\nu$

Z = 92

The K_α X-rays correspond to the transition from the L-shell to the K-shell which takes place when one of the K-electrons has been removed. The L-shell electron will see an effective charge (Z-1). Then:

$$h\nu = chR(Z-1)^2\left(\frac{1}{n_f^2} - \frac{1}{n_i^2}\right) = chR(Z-1)^2\left(\frac{1}{1^2} - \frac{1}{2^2}\right)$$

or:

$$h\nu = \frac{3}{4}\,chR(Z-1)^2$$

Numerically, using CGS units:

$$h\nu = \frac{3}{4}\;\frac{3 \times 10^{10}(\text{cm/sec}) \times 6.626 \times 10^{-27}(\text{erg-sec}) \times 109740(\text{cm}^{-1}) \times (91)^2}{1.6 \times 10^{-9}\ (\text{erg/Kev})}$$

and: $\boxed{h\nu = 84.6 \text{ Kev}}$ Answer

Photons of larger energies are observed because uranium is naturally radioactive, and will decay emitting alpha particles and gamma photons. The origin of these photons is thus the nucleus.

****

62

In the Stern-Gerlach experiment, silver atoms evaporated in an oven at temperature T are collimated by passingthem through two narrow slits, and then enter an inhomogeneous magnetic field of constant gradient, $\partial B/\partial z$.The magnetic field extends for a length a, after which the beam travels a further distance b before striking a glass plate.

(a) Assuming that all the atoms have the same speed, as defined by $Mv^2/2 = 3kT/2$, show that the separation of the two lines on the plate is given by:

$$d = \frac{ae\hbar}{3mkT}\ \frac{\partial B}{\partial z}\ (\frac{a}{2} + b)$$

where m is the mass of the electron.

(b) Would this result change if different isotopes of silver were in the beam ?

****

(a) The horizontal velocity is constant, and given by:

$$\frac{Mv_x^2}{2} = \frac{3}{2}\,kT\ ,\ \text{or:}\quad v_x = \sqrt{\frac{3kT}{M}}$$

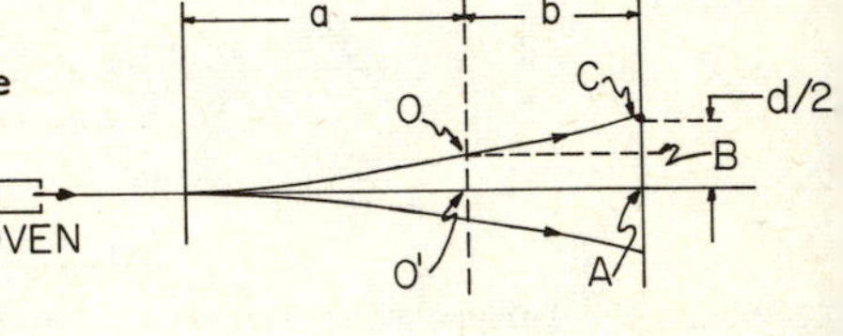

Consequently, the times required to travel the distances a and b are respectively:

$$t_a = \frac{a}{v_x} = \frac{a\sqrt{M}}{\sqrt{3kT}}$$

$$t_b = \frac{b}{v_x} = \frac{b\sqrt{M}}{\sqrt{3kT}}$$

The silver atoms have a magnetic moment which is due exclusively to the spin of the 5s electron, and in the magnetic field will be acted upon by a force of magnitude:

$$F_z = m_s\,\frac{e\hbar}{m}\,\frac{\partial B}{\partial z}$$

Since the spin magnetic quantum number m_s can take values $+\frac{1}{2}$ and $-\frac{1}{2}$, the force will be either up or down, and the atoms will be deviated accordingly. Taking, for instance, $m_s = 1/2$,

$$F_z = \frac{e\hbar}{2m}\,\frac{\partial B}{\partial z} = Ma_z$$

indicating that the vertical acceleration is constant. Then:

$$v_z = a_z t \qquad (1)$$

and:

$$z = \frac{1}{2}\,a_z t^2 \qquad (2)$$

At point 0 :

$$v_z = a_z t_a = \frac{1}{M}\,\frac{e\hbar}{2m}\,\frac{\partial B}{\partial z}\,\frac{a\sqrt{M}}{\sqrt{3kT}}$$

and since this velocity is thereon constant:

$$BC = v_z t_b = \frac{1}{M}\,\frac{e\hbar}{2m}\,\frac{\partial B}{\partial z}\,\frac{a\sqrt{M}}{\sqrt{3kT}}\,\frac{b\sqrt{M}}{\sqrt{3kT}} = \frac{e\hbar}{2m}\,\frac{\partial B}{\partial z}\,\frac{ab}{3kT} \qquad (3)$$

The distance AB is given by equation (2) for $t = t_a$:

$$AB = \frac{1}{2} a_z t_a^2 = \frac{1}{2} \frac{1}{M} \frac{e\hbar}{2m} \frac{\partial B}{\partial z} \frac{a^2 M}{3kT} = \frac{e\hbar}{2m} \frac{\partial B}{\partial z} \frac{a^2}{6kT} \qquad (4)$$

The total deviation is obtained by adding (3) and (4):

$$\frac{d}{2} = \frac{e\hbar}{2m} \frac{\partial B}{\partial z} \frac{ab}{3kT} + \frac{e\hbar}{2m} \frac{\partial B}{\partial z} \frac{a^2}{6kT}$$

or, finally:

$$\boxed{d = \frac{ae\hbar}{3mkT} \frac{\partial B}{\partial z} \left(\frac{a}{2} + b\right)} \qquad \text{Q.E.D.}$$

(b) Since the atomic mass does not appear in the above formula, clearly the presence of different isotopes will not change the result. In fact, atoms of other elements, like for example potassium, would give the same separation, provided the magnetic moment is due to a single electron.

****

63

Silver atoms are heated to a temperature of 452°C and the beam is collimated and passed through an inhomogeneous magnetic field of constant gradient 100 weber/m^2 per centimeter for a distance of 3 cm. The beam then travels a distance of 8.5 cm beyond the end of the magnetic field and strikes a photographic plate. Calculate the maximum separation between the two lines on the plate.

****

This problem is just a numerical application of the formula developed in the previous problem:

$$d = \frac{2a}{3kT}\frac{e\hbar}{2m}\frac{\partial B}{\partial z}\left(\frac{a}{2} + b\right) \qquad (1)$$

Notice that the formula is written in a slightly different form, to take advantage of the fact that the combination:

$$\frac{e\hbar}{2m} = \beta = \text{Bohr magneton}$$

has a well known numerical value:

$$\beta = 9.274 \times 10^{-21}\ \frac{\text{erg}}{\text{gauss}}$$

Also:

$$\left.\begin{array}{l} a = 3 \text{ cm} \\ b = 8.5 \text{ cm} \end{array}\right\} \therefore \frac{a}{2} + b = 1.5 + 8.5 = 10 \text{ cm}$$

$$T = 452 + 273 = 725°K$$

and (1) becomes:

$$d = \frac{2 \times 3\ (\text{cm}) \times 9.274 \times 10^{-21}(\text{erg/gauss}) \times 10^{6}(\text{gauss/cm}) \times 10\ (\text{cm})}{3 \times 1.38 \times 10^{-16}(\text{erg/°K}) \times 725(°\text{K})}$$

or:

$$\boxed{d = 1.85 \text{ cm}}$$ Answer

****

64

(a) Compute the Landé g-factors for the transition: $^2P_{1/2} \rightarrow {}^2S_{1/2}$.

(b) In a weak external magnetic field the levels will split. How many sublevels will there be for each level ? Draw an energy level scheme and indicate all the allowed transitions between sublevels of the upper state and sublevels of the lower state.

(c) How many spectral lines will there be instead of the one seen with no magnetic field ?

****

(a) First, one recalls that the $^2S_{1/2}$ state is not really a doublet, since for $L = 0$ the only value of J is $1/2$, but that to retain the symmetry of the notation it is conventional to write the subscript 2.

The Landé g-factor is:

$$g = 1 + \frac{J(J+1) + S(S+1) - L(L+1)}{2J(J+1)} = 1 + \frac{1}{2} + \frac{1}{2}\frac{S(S+1)}{J(J+1)} - \frac{1}{2}\frac{L(L+1)}{J(J+1)}$$

and for: $L = 0$, $J = S = \frac{1}{2}$, one has: $\boxed{g = 2}$ Ans.(a)

The $^2P_{1/2}$ state is a doublet and: $L = 1$, $J = \frac{1}{2}$

Since the number of values of J is 2, then: $2 = 2S + 1$, or: $S = \frac{1}{2}$.

Then:

$$g = 1 + \frac{1}{2} + \frac{1}{2} - \frac{4}{3} \qquad \text{or:} \qquad \boxed{g = \frac{2}{3}}$$ Ans.(a)

(b) In a weak magnetic field the energy levels will split into as many levels as the possible values of the magnetic quantum number m_J, which are $(2J+1)$, or, for $J = 1/2$, just two values:

$$m_J = \pm\frac{1}{2}$$

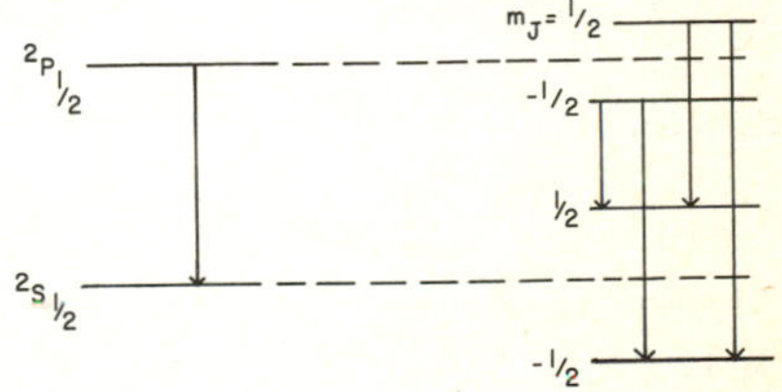

Hence, each level will split into two, as shown in the diagram at right. The allowed transitions must obey the selection rule:

$$\boxed{\Delta m_J = \pm 1, 0}$$

and consequently there will be 4 allowed transitions as indicated.

(c) The separation of the split levels is not the same, since it depends on the value of the g-factor. Thus, each allowed transition has different ΔE, and will give one spectral line, i.e., 4 lines will replace the original line observed with no magnetic field.

Notice that for $n = 3$, the transition $^2P_{1/2} \rightarrow {}^2S_{1/2}$ corresponds to the D_1 line of sodium, which indeed exhibits this anomalous Zeeman effect.

****

65

A beam of neutral boron atoms moving in the x-direction with a speed of 10^5 m/sec enters a region where the magnetic field has a gradient in the y-direction,

$$\frac{dB}{dy} = \frac{1}{2} \; \frac{\text{wb/m}^2}{\text{cm}}$$

but is independent of x. The deflections of the atomic beam are very small, and along the path of the beam the field is $B = 1$ wb/m^2. The field extends for 20 cm in the x-direction, and after leaving it the atoms travel a further distance of 90 cm before striking a screen. This is an atomic problem, and all contributions from the nucleus are to be neglected.

(a) Determine all the possible values of the total angular momentum of the boron atoms.

(b) Determine all the possible values of the Larmor precession frequency.

(c) Find an expression, in terms of the different parameters of the problem, giving the various deflections which will be observed on the screen as a consequence of the magnetic dipole moments of the boron atoms. Calculate those deflections in cm.

****

(a) The electronic configuration of boron (Z = 5) is:

$$1s^2 2s^2 2p^1$$

The K-shell and the 2s subshell are complete, and do not contribute to the angular momentum, which is entirely due to the orbital motion and spin of the 2p electron. This electron has:

$$\ell = 1 \,, \qquad \text{and:} \qquad s = \frac{1}{2}$$

and thus, for the atom: $\quad L = 1 \,, \qquad \text{and:} \qquad S = \frac{1}{2}$

The total angular momentum J will take 2S+1 = 2 values, which are:

$$J = L + S = 1 + \frac{1}{2} = \frac{3}{2}$$

$$J = L - S = 1 - \frac{1}{2} = \frac{1}{2}$$

Thus, the lowest energy state is a P doublet:

$$^2P_{3/2} \,, \; ^2P_{1/2}$$

The total angular momentum of the atoms takes the values:

$$p'_J = \sqrt{J(J+1)}\,\hbar = \sqrt{\frac{3}{2}\left(\frac{3}{2}+1\right)}\,\hbar = \frac{\sqrt{15}}{2}\,\hbar$$

$$p''_J = \sqrt{J(J+1)}\,\hbar = \sqrt{\frac{1}{2}\left(\frac{1}{2}+1\right)}\,\hbar = \frac{\sqrt{3}}{2}\,\hbar$$

or, since: $\quad \hbar = \frac{h}{2\pi} = 1.054 \times 10^{-34}$ joule-sec

$$\boxed{p'_J = 2.04 \times 10^{-34} \text{ joule-sec}} \,, \qquad \boxed{p''_J = 9.11 \times 10^{-35} \text{ joule-sec}} \qquad \underline{\text{Ans.}}$$

(b) Classical electromagnetic theory shows that if the magnetic moment $\underline{\mu}$ of a body is proportional to its vector angular momentum $\underline{p}$, when the body is in a magnetic field $\underline{B}$ the vector $\underline{p}$ will precess about the direction of $\underline{B}$ with the Larmor frequency:

$$\omega_L = \frac{\mu}{p} B \qquad \frac{\text{rad}}{\text{sec}} \qquad (1)\dagger$$

In the present case: $\mu_J = \frac{e}{2m} g \, p_J$

where the Landé g-factor is given by the expression:††

$$g = 1 + \frac{J(J+1) + S(S+1) - L(L+1)}{2J(J+1)}$$

Equation (1) becomes: $\omega_L = \frac{\mu_J}{p_J} B = \frac{e}{2m} g \, B$

Numerically: for $J = \frac{3}{2}$, $g = \frac{4}{3}$

for $J = \frac{1}{2}$, $g = \frac{2}{3}$

Also: $\frac{e}{m} = 1.75879 \times 10^{11} \frac{\text{coul}}{\text{Kg}}$, and: $B = 1 \frac{\text{weber}}{m^2}$

$$\therefore \quad \omega_L' = \frac{1}{2} \times 1.7588 \times 10^{11} \times \frac{4}{3} \times 1 = \boxed{1.172 \times 10^{11} \frac{\text{rad}}{\text{sec}}} \qquad \text{Ans.(b)}$$

Likewise: $$\omega_L'' = \frac{1}{2} \times 1.7588 \times 10^{11} \times \frac{2}{3} \times 1 = \boxed{5.86 \times 10^{10} \frac{\text{rad}}{\text{sec}}} \qquad \text{Ans.(b)}$$

(c) The potential energy of a magnetic dipole moment in a magnetic field $\underline{B}$ is:

$$V = - \underline{\mu} \cdot \underline{B}$$

Hence, the force is: $$F = - \frac{dV}{dy} = \underline{\mu} \cdot \frac{d\underline{B}}{dy} = \mu \frac{dB}{dy} \cos \theta \qquad (2)$$

where: $\mu = \mu_J = \frac{eg}{2m} p_J = \frac{eg}{2m} \sqrt{J(J+1)} \, \hbar$

$\cos \theta = m_J / \sqrt{J(J+1)}$

The magnetic quantum number m_J takes $(2J+1)$values; for $J = \frac{3}{2}$:

$$m_J = \frac{3}{2}, \frac{1}{2} , - \frac{1}{2} , - \frac{3}{2}$$

and for $J = \frac{1}{2}$: $m_J = \frac{1}{2} , - \frac{1}{2}$

Equation (2) is written then:

$$F = \frac{eg}{2m} \sqrt{J(J+1)} \, \hbar \frac{dB}{dy} \frac{m_J}{\sqrt{J(J+1)}} = m_J \frac{e\hbar}{2m} g \frac{dB}{dy}$$

But: $\frac{e\hbar}{2m} = \beta = \text{Bohr magneton} = 9.273 \times 10^{-21}$ erg/gauss

and the force is finally expressed: $$\boxed{F = m_J \beta g \frac{dB}{dy}} \qquad (3)$$

---

†) See, for example: Richtmyer, Kennard and Lauritsen, Introduction to Modern Physics, McGraw-Hill, 1955, p.285.

††) Op.cit., p.292.

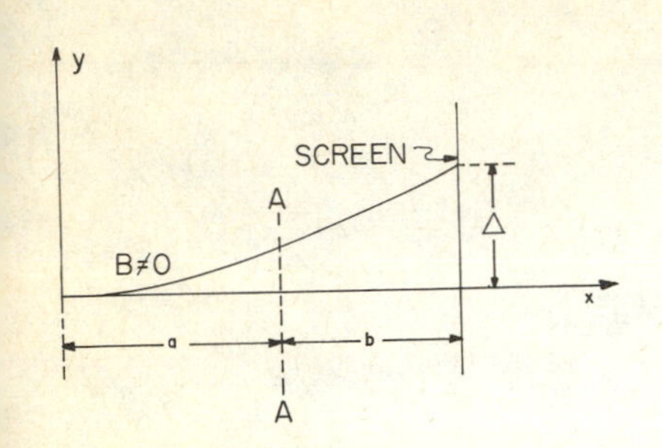

Let: a = distance atoms move within the magnetic field
b = distance from end of field to screen

The transit times are:

$$t_a = \frac{a}{v}, \quad \text{and:} \quad t_b = \frac{b}{v}$$

The vertical acceleration, a_y, is constant:

$$a_y = \frac{F}{M}$$

and:
$$y = \frac{1}{2} a_y t^2 = \frac{1}{2} \frac{F}{M} t^2$$

$$v_y = a_y t = \frac{F}{M} t$$

which, at the moment the atoms leave the region of the magnetic field take the values:

$$y_1 = \frac{1}{2} \frac{F}{M} t_a^2 = \frac{1}{2} \frac{F}{M} \frac{a^2}{v^2}$$

$$v_{y_1} = \frac{F}{M} \frac{a}{v}$$

From AA to the screen the atoms have constant velocity and:

$$y = v_{y_1} t + y_1$$

which at $t = t_b$ takes the value:

$$\Delta = \frac{F}{M} \frac{a}{v} \frac{b}{v} + \frac{1}{2} \frac{F}{M} \frac{a^2}{v^2} = \frac{Fa}{Mv^2} [b + \frac{a}{2}]$$

and replacing F by its expression (3):

$$\boxed{\Delta = \frac{a\beta}{Mv^2} m_J g \frac{dB}{dy} [b + \frac{a}{2}]} \qquad (4)$$

Answer (c)

For $J = 3/2$, $g = 4/3$, and m_J takes 4 values, while for $J = 1/2$, $g = 2/3$, and m_J takes two values. Thus, there will be a total of six deflections on the screen.

The actual values of the deflections can be better computed if all the constants are grouped into one:

$$K = \frac{a\beta}{Mv^2} \frac{dB}{dy} [b + \frac{a}{2}] \qquad (5)$$

Here:
$$Mv^2 = 10.82 \text{ (amu)} \times 1.66 \times 10^{-24} (\frac{gm}{amu}) \times 10^{-3} (\frac{Kg}{gm}) \times (10^5)^2 (\frac{m}{sec})^2$$

$$= 1.8 \times 10^{-16} \text{ joule} = 1.8 \times 10^{-9} \text{ erg}$$

$$b + \frac{a}{2} = 90 + \frac{20}{2} = 100 \text{ cm}$$

and K is:
$$K = \frac{20 (cm) \times 9.273 \times 10^{-21} (erg/gauss) \times 5000 (gauss/cm) x100 (c}{1.8 \times 10^{-9} (erg)}$$

or: $K = 5.15 \times 10^{-5}$ cm

Equation (4) becomes then: $\Delta = 5.15 \times 10^{-5}\ m_J g$ cm

and the deviations are easily calculated. The results are tabulated below:

$J = \frac{3}{2}$, $g = \frac{4}{3}$	$m_J = \pm \frac{3}{2}$	$\Delta = \pm 1.03 \times 10^{-4}$ cm
	$m_J = \pm \frac{1}{2}$	$\Delta = \pm 0.343 \times 10^{-4}$ cm
$J = \frac{1}{2}$, $g = \frac{2}{3}$	$m_J = \pm \frac{1}{2}$	$\Delta = \pm 0.172 \times 10^{-4}$ cm

****

66

A single alpha particle having a kinetic energy of 1 Mev strikes normally a thin foil of copper of thickness 0.01 mm and density 8.8 grams/cm^3. Calculate the probability that the alpha particle will be deflected through an angle which is greater than 30° and less than 60°.

Answer: 5%

****

6

NUCLEAR PHYSICS

(a) The nucleus and radiactive decay.

67 Using the uncertainty principle, show that an electron cannot exist within a nucleus.

****

In terms of position and momentum, the uncertainty principle is written:

$$\Delta x \, \Delta p \sim \hbar \qquad (1)$$

If the electron is inside the nucleus, the uncertainty in position will be equal to the nuclear diameter. But the nuclear radius is:

$$R = 1.3 \times 10^{-13} \sqrt[3]{A} \quad \text{cm}$$

and taking, say, A = 57:

$$R = 1.3 \times 10^{-13} \sqrt[3]{57} = 5 \times 10^{-13} \text{ cm}$$

From (1):

$$\Delta p \sim \frac{\hbar}{2R}$$

and since this electron is extreme relativistic its energy can be as large as:

$$\Delta E = c \, \Delta p \sim \frac{\hbar c}{2R}$$

Numerically:

$$\Delta E = \frac{1.054 \times 10^{-27} \text{ (erg-sec)} \times 3 \times 10^{10}}{2 \times 5 \times 10^{-13} \text{(cm)} \times 1.6 \times 10^{-6} \text{ (erg/Mev)}} = 19.75 \text{ Mev}$$

and since $\Delta E \sim K$, the kinetic energy of the electrons in the nucleus would be:

$$\boxed{K \sim 20 \text{ Mev}}$$

which is too large when one considers that the average binding energy for nucleons is about 8 Mev. Thus, it seems unreasonable that electrons should exist within the nucleus with such kinetic energies. This argument, among others, led to the abandonning of the proton-electron model of the nucleus.

****

68 Estimate the density of nuclear matter.

****

The nuclear density is given by the ratio:

$$\rho = \frac{M}{V} \tag{1}$$

where M represents the nuclear mass and V the nuclear volume. The value of M, in grams, is:

$$M = (\text{nuclear mass in amu}) \times \text{conversion factor (gm/amu)}$$

But the atomic mass can be approximated by the mass number A, and:

$$M = Af \tag{2}$$

To calculate V one needs a value for the nuclear radius R, which is given by the expression:

$$R = r_o A^{1/3}$$

r_o is a constant, ranging from 1.2 to 1.5 fermis ($1 \text{ f} = 10^{-13}$ cm), depending on the experimental procedure followed for its determination.†
For this estimate, it will be assumed that:

$$r_o = 1.2 \text{ fermis}$$

Then:

$$V = \frac{4}{3}\pi R^3 = \frac{4}{3}\pi r_o^3 A \tag{3}$$

and since r_o can be interpreted as the radius of a single nucleon, the nuclear volume is:

$$V = [\text{volume of one nucleon}][\text{number of nucleons}]$$

which indicates that the nuclear density is the same for all nuclei. To compute its numerical value, one substitutes (2) and (3) into (1):

$$\rho = \frac{Af}{\frac{4}{3}\pi r_o^3 A} = \frac{3f}{4\pi r_o^3}$$

Numerically:

$$\rho = \frac{3 \times 1.66 \times 10^{-24}}{4\pi \times (1.2)^3 \times 10^{-39}}$$

$$\therefore \quad \boxed{\rho = 2.3 \times 10^{14} \text{ gm/cm}^3}$$ Answer

****

(†) A good discussion of this point can be found in C.M.H. Smith, Nuclear Physics, Pergamon Press, 1966, Chapter 9.

69

The present scale of atomic masses is based on the arbitrary assignment of 12.00000 amu to C^{12}.† The old system was based on O^{16} = 16.00000 amu. In the old system, the packing fraction of C^{12} was 3.17×10^{-4}. The mass of Al^{27} is 26.9815 amu in the new system. What was its mass in the previous system ?

****

Let: M = atomic mass amu = units based on O^{16}

A = mass number amu* = units based on C^{12}

The mass defect is defined to be: $\Delta = M - A$

and the packing fraction: $F = \frac{\Delta}{A} = \frac{M-A}{A}$

and from here:

$$M = A(1+F)$$

For C^{12}: $M = 12(1 + 3.17 \times 10^{-4})$ amu = 12.00000 amu*

or: 1 amu* $= (1 + 3.17 \times 10^{-4})$ amu

Then, for Al^{27}:

$$M = 26.9815 \text{ amu*} = 26.9815\,(1 + 3.17 \times 10^{-4}) \text{ amu}$$
$$= 26.9815 + 8.56 \times 10^{-3}$$

or: $\boxed{Al^{27} = 26.99006 \text{ amu}}$ Answer

Notice that the listed value for the mass of Al^{27} based on O^{16} is 26.990081 amu.

****

(†) As recommended by the International Commission of Atomic Weights.

70

Carbon 10 is unstable, and decays by positron emission. Coincidence-counter experiments show that in about 2% of the disintegrations two gamma rays of energies 1.02 and 0.72 Mev respectively are also emitted, while in 98% of the cases only the 0.72 Mev gamma ray is present.

(a) Show a disintegration scheme for C^{10}.

(b) Calculate the disintegration energy, Q, and determine the end-point energy of the positrons.

****

(a) The nuclear reaction is written:

$$_6C^{10} \rightarrow {_5B^{10}} + \beta^+ + \nu + Q$$

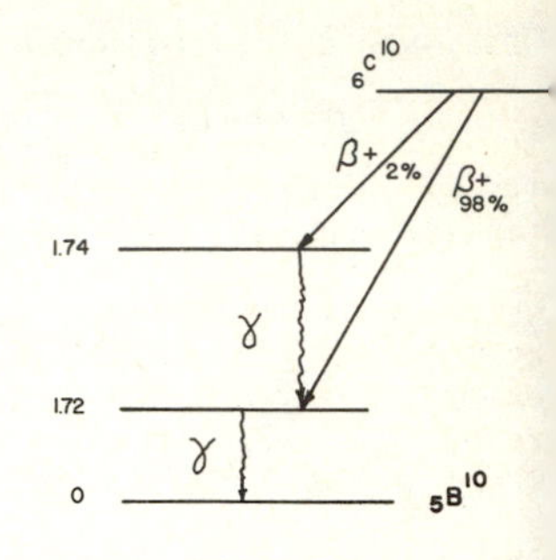

Since the 0.72 Mev gamma follows all disintegrations, it indicates that the nucleus of B^{10} is left in an excited state, passing to the ground state by emission of this gamma. When two gamma rays are ejected, the B^{10} nucleus is at a higher energy state, namely 0.72 + 1.02 = 1.74 Mev, and decays to the ground state by cascade emission of the two photons. The disintegration scheme is then as shown at right.

(b) By adding the proper number of electron masses to both sides of equation (1), one obtains a relation between the atomic masses and:

$$Q = {_6M^{10}} - ({_5M^{10}} + 2m_e) \qquad \text{amu}$$

Using atomic masses based on oxygen 16:†

$$Q = 10.020240 - (10.016119 + 0.001098) = 0.003013 \text{ amu}$$

$$= 0.003013 \times 931.5 = 2.80 \text{ Mev}$$

Hence: $\boxed{Q = 2.80 \text{ Mev}}$ <u>Answer</u>

The end-point energy of the positrons is the difference between the disintegration energy and the energy of the excited nucleus, i.e, the energy carried away by the gamma photons. Thus, in 98% of the cases:

$$E_{\beta+} = 2.80 - 0.72$$

or: $\boxed{E_{\beta+} = 2.08 \text{ Mev}}$ <u>Answer</u>

which compares well with the experimental value of 2.1 Mev.††

In the remaining 2% of the cases, the end-point energy of the positron is:

$$E_+ = 2.80 - 1.74 = \boxed{1.06 \text{ Mev}}.$$ <u>Answer</u>

****

(†) Radiological Health Handbook, U.S. Dept. of Health, Education and Welfare, (Clearinghouse for Federal Scientific and Technical Information, Sprigfield, Va.), PB 121784R, 1960.

(††) Op.cit., p.223

71

N^{13} is unstable and decays by positron emission. The maximum energy of the positrons is observed to be 1.20 Mev. (a) Show a disintegration scheme for the β^+ decay of N^{13}. (b) From the above reaction, calculate the atomic mass of N^{13}. Use atomic masses based on C^{12}.

****

(a) Positron emission occurs when a proton is transformed into a neutron inside the nucleus:

$$p \rightarrow n + \beta^+ + \nu$$

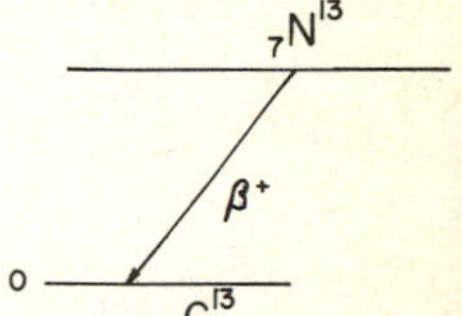

The nuclear reaction is expressed:

$$_7N^{13} \rightarrow {}_6C^{13} + \beta^+ + \nu + Q \qquad (1)$$

where Q, the disintegration energy, is carried away by the neutrino and the positron. The maximum positron energy corresponds to the case when all the energy released is carried away by the β^+, so in this case Q = 1.20 Mev . The disintegration scheme is as shown.

(b) Since the masses usually tabulated are atomic masses, one can transform equation (1) into an atomic equation by adding to both sides the proper number of electronic masses, in this case, seven:

$$_7M^{13} = {}_6M^{13} + 2m_e + Q \qquad (2)$$

Now, the atomic masses based on C^{12} are:

$$_6M^{13} = 13.003354 \text{ amu}$$
$$m_e = 0.0005485 \text{ amu}$$

Also: $$Q = \frac{1.20 \text{ (Mev)}}{931.5 \text{ (Mev/amu)}} = 0.001288 \text{ amu}$$

Equation (2) yields:

$$_7M^{13} = 13.003354 + 0.001097 + 0.001288$$

or: $$\boxed{_7M^{13} = 13.005739 \text{ amu}}$$

while the listed atomic mass of $_7N^{13}$ is 13.005738 amu.

****

72

Discuss the possibilities of isobaric transformation for the isobaric pair Mn^{53} and Cr^{53}, and state which one of them is energetically possible. Use the following atomic masses:†

$$Mn^{53} = 52.958100 \text{ amu}, \qquad Cr^{53} = 52.957460 \text{ amu}.$$

****

First one observes that the mass of $_{24}Cr^{53}$ is not larger than the mass of $_{25}Mn^{53}$, so that negative beta decay is not possible.

Now, Mn^{53} can decay into Cr^{53} by either positron emission:

$$_{25}Mn^{53} \rightarrow {}_{24}Cr^{53} + \beta^{+} + \nu \qquad (1)$$

or by K-capture:

$$_{25}Mn^{53} + \beta^{-} \rightarrow {}_{24}Cr^{53} + \nu \qquad (2)$$

For the reaction (1) to be possible, the Q-value must be:

$$Q = (\text{initial mass}) - (\text{final mass} + 2m_o) \geqslant 0$$

Here: $Q = 52.9581 - (52.95746 + 2 \times 0.00055) = 52.9581 - 52.9585 < 0$

and β^{+} decay is not possible. Hence, K-capture is the only energetically possible transformation. In fact, the disintegration scheme of Mn^{53} is as shown at right.††

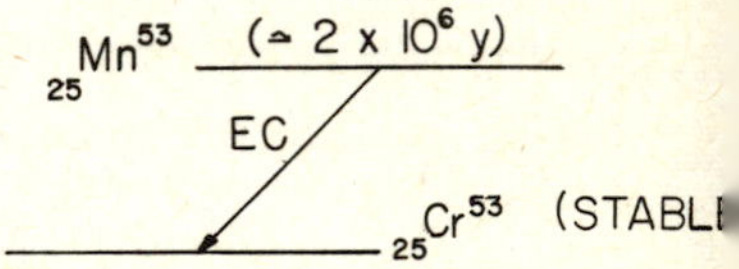

$_{25}Mn^{53}$ ($\simeq 2 \times 10^{6}$ y)
EC
$_{25}Cr^{53}$ (STABLE)

****

(†) Based on oxygen 16. See Radiological Health Handbook, Dept.of Health, Education and Welfare, 1960, PB121784R.

(††) op.cit.,page 245.

73 Consider the isobars ${}_{18}Ar^{37}$ and ${}_{17}Cl^{37}$, with atomic masses 36.978416 and 36.977540 amu respectively.†

(a) Determine whether it is energetically possible for the transmutation of Ar^{37} to Cl^{37} to occur by electron capture or positron emission.
(b) Calculate the disintegration energy Q in Mev.
(c) Using conservation of energy and momentum, show that the kinetic energy of the recoil nucleus is expressed:

$$E_r = \frac{Q^2}{2Mc^2}$$

where M is the mass of the daughter nucleus.
(d) Compute the velocity of the recoil nucleus in m/sec.

****

(a) Decay by positron emission is possible only if the atomic mass of the parent exceeds the atomic mass of the daughter by two electron masses.
For the present case:

parent, Ar^{37}:	36.978416
daughter, Cl^{37}:	36.977540
	Δm = .000876 amu

which is less than two electron masses. Hence, positron emission is not possible, and the decay will occur by electron capture.

(b) For EC the disintegration energy is just the difference between the atomic masses of parent and daughter. Using the conversion factor, 1 amu = 931.5 Mev, one has:

$$Q = 0.000876 \times 931.5 \qquad \text{or:} \qquad \boxed{Q = 0.815 \text{ Mev}} \qquad \text{Ans.(b)}$$

(c) When an electron is captured by the nucleus, a proton is transformed into a neutron, while a neutrino is ejected:

$$p + \beta^- \rightarrow n + \nu$$

The disintegration energy is shared by the recoil nucleus and the neutrino:

$$Q = E_r + E_\nu = \frac{p^2}{2M} + p_\nu c \qquad (1)$$

By conservation of linear momentum: $p = p_\nu$ (2)

Eliminating p_ν between (1) and (2), one has:

$$Q = \frac{p^2c^2}{2Mc^2} + pc$$

$$\text{or:} \qquad \frac{(pc)^2}{2Mc} + pc - Q = 0$$

Solving for pc:

$$pc = \frac{-1 \pm \sqrt{1 + \frac{4Q}{2Mc^2}}}{2/2Mc^2} = Mc^2[-1 \pm \sqrt{1 + \frac{4Q}{2Mc^2}}]$$

Since pc is essentially positive, only the + sign has physical significance. Moreover, always $4Q << 2Mc^2$, i.e.,

$$\frac{4Q}{2Mc^2} << 1$$

†) Op.cit.

One can use then the approximation:

$$(1+x)^{1/2} = 1 + \frac{x}{2} \qquad \text{where: } x = \frac{4Q}{2Mc^2}$$

to get:

$$pc = Mc^2[-1 + 1 + \frac{1}{2}\frac{4Q}{2Mc^2}] \equiv Q$$

Hence:

$$E_r = \frac{p^2}{2M} = \frac{p^2c^2}{2Mc^2}$$

or:

$$\boxed{E_r = \frac{Q^2}{2Mc^2}}$$

Q.E.D.

(d) The velocity of the recoil nucleus is:

$$v = \frac{p}{M} = \frac{pc}{Mc} = \frac{Q}{Mc}$$

Numerically, using MKS units:

$$v = \frac{0.815 \text{ (Mev)} \times 1.6 \times 10^{-13} \text{(joules/Mev)}}{36.977 \text{(amu)} \times 1.66 \times 10^{-27} \text{(Kg/amu)} \times 3 \times 10^{8} \text{(m/sec)}}$$

or:

$$\boxed{v = 7.08 \times 10^{3} \text{ m/sec}}$$

Ans.(d)

****

74 Consider the mirror nuclei $_6C^{11}$ and $_5B^{11}$ which have atomic masses 11.014922 amu and 11.012795 amu respectively.
(a) Complete the equation: $_6C^{11} \rightarrow {}_5B^{11} + \ldots$
Is this nuclear disintegration possible ?
(b) If yes, what would be the energy available ?
(c) What would be the average energy of the particle ejected ?

****

(a) Since the atomic number Z has been increased by one unit, the disintegration has taken place by positron emission or by electron capture. In the first case, the nuclear reaction is written:

$$_6C^{11} \rightarrow {}_5B^{11} + \beta^+ + \nu + Q$$

where Q, the energy resulting from the process, has to be positive for the reaction to occur. Adding 6 electron masses to each side of the equation, we rewrite it in terms of atomic masses:

$$_6M^{11} \rightarrow {}_5M^{11} + 2m_e + Q$$

or:

$$Q = {}_6M^{11} - ({}_5M^{11} + 2m_e) = 11.014922 - (11.012795 + 0.001098)$$

$$= 0.001029 \text{ amu} = 0.001029 \times 931.5 \text{ Mev} = 0.956 \text{ Mev}$$

Thus, $Q > 0$, indicating that the disintegration is possible. Notice that electron capture is also energetically possible, and will also occur with a very small probability (less than 1%).

(b) As calculated in part (a), the disintegration energy is:

$$\boxed{Q = 0.956 \text{ Mev}}$$ Ans.(b)

which is the energy available, and will be carried away by the positron and the neutrino. The positron energy will have then a maximum value:

$$E_{max} = 0.956 \text{ Mev} \qquad \text{(end-point energy)}$$

(c) The positrons resulting from the disintegration will have a distribution in energy up to the value E_{max}. A good approximation for the average positron energy is 1/3 of the maximum value:

$$\boxed{\langle E \rangle = \frac{1}{3} E_{max} = 0.318 \text{ Mev}}$$ Ans.(c)

****

75 The activity of a certain radioisotope is observed to decrease by a factor of 8 in 36 hours. (a) What is the half-life of the isotope ? (b) What is the disintegration constant ? (c) If the present activity of the sample is 400 millicuries, what will be the activity after 12 hours ?

****

(a) This is a very simple problem that requires practically no calculation. By definition of half-life, the activity will be reduced by a factor of 1/2 in one half-life, and by a factor of $1/2^n$ in n half-lives. Here, n = 3, and

$$3T_{1/2} = 36 \text{ hr} \qquad \therefore \qquad \boxed{T_{1/2} = 12 \text{ hours}}$$ Ans.(a)

(b) The disintegration constant is:

$$\lambda = \frac{\ln 2}{T_{1/2}} = \frac{0.693}{12} \qquad \text{or:} \qquad \boxed{\lambda = 0.0577 \text{ hr}^{-1}}$$ Ans.(b)

(c) Since the half-life is 12 hours, the activity will be reduced by one-half:

$$\boxed{A = 200 \text{ mc}}$$ Ans.(c)

****

76

The half-life of $_{90}Th^{234}$ is 24.1 days. How long after a sample of this isotope has been isolated will it take for 90% of it to change to $_{91}Pa^{234}$?

****

Radioactive decays follows the exponential law:

$$\boxed{N = N_o e^{-\lambda t}}$$

where: N = number of atoms present at time t

λ = disintegration constant = $\ln 2/T_{1/2}$

In the present case, $N = 0.1\ N_o$ and:

$$0.1\ N_o = N_o \exp\left(-\frac{0.693\ t}{24.1}\right)$$

or:

$$\ln 10 = \frac{0.693\ t}{24.1}$$

Solving for t:

$$t = \frac{24.1 \times 2.3}{0.693} = \boxed{80 \text{ days}}$$

Answer

****

77

Careful radioactivity measurements on a sample of ${}_6C^{14}$ indicate that there are 10^5 carbon disintegrations per second. The half-life of C^{14} is 5568 years. Calculate the size of the sample in grams.

****

The activity of a radioactive source is:

$$A = \lambda N \qquad \text{dis/sec} \qquad (1)$$

where λ, the disintegration constant, is given by the relation:

$$\lambda = \frac{\ln 2}{T_{1/2}}$$

and N, the number of atoms in the sample, is:

$$N = \frac{N_o W}{M} \qquad \text{where: } \begin{cases} N_o = \text{Avogadro's number} \\ W = \text{mass of sample, gm} \\ M = \text{atomic mass} \end{cases}$$

Combining all these formulas, equation (1) becomes:

$$A = \frac{\ln 2}{T_{1/2}} \frac{N_o W}{M}$$

and solving for W:

$$W = \frac{MAT_{1/2}}{(\ln 2) N_o} \qquad \text{grams}$$

Numerically:

$$W = \frac{10^5 \times 14 \times 5568 \times 3.157 \times 10^7}{0.693 \times 6.023 \times 10^{23}}$$

or:

$$\boxed{W = 5.9 \times 10^{-7} \text{ grams}}$$

Answer

****

78 A pulse of 10^8 thermal neutrons (i.e., v = 2200 m/sec) move through a vacuum for a distance of 11 meters before striking a target. If the half-life of free neutrons is 12 minutes, determine how many neutrons will spontaneously decay while traveling toward the target.

****

The decay of the neutrons will follow the exponential law:

$$N = N_o e^{-\lambda t}$$

where t, the time required for the neutrons to reach the target, is:

$$t = \frac{d}{v} = \frac{11 \text{ (m)}}{2200 \text{(m/sec)}} = 5 \times 10^{-3} \text{ sec}$$

The number of neutrons that decay in that time is:

$$\Delta N = N_o - N = N_o - N_o e^{-\lambda t} = N_o(1 - e^{-\lambda t}) \qquad (1)$$

But: $$\lambda t = \frac{5 \times 10^{-3} \text{ (sec)}}{12 \text{ (min)} \times 60 \text{ (sec/min)}} = 6.95 \times 10^{-6}$$

which is a very small number. Using the approximation:

$$e^{-\lambda t} \simeq 1 - \lambda t$$

equation (1) becomes:

$$\Delta N = N_o(1 - 1 + \lambda t) = N_o \lambda t$$

or, numerically: $$\Delta N = 10^8 \times 6.95 \times 10^{-6}$$

or: $$\boxed{\Delta N = 695 \text{ neutrons}}$$ Answer

****

79

The present-day isotopic abundance ratio of U^{235} to U^{238} is 0.0072. Assuming that all the uranium of the galaxy was formed at the same time, and that the initial amounts of U^{235} and U^{238} were the same, estimate the age of the galaxy.

****

The half-lives of the uranium isotopes are:†

$$U^{238} : \quad T_1 = 4.51 \times 10^9 \text{ years}$$
$$U^{235} : \quad T_2 = 7.13 \times 10^8 \text{ "}$$

Let N_o be the initial number of atoms of either isotope. After a certain time t, the number of atoms of U^{238} is:

$$N_1 = N_o \exp(-\frac{0.693\,t}{T_1})$$

while the number of atoms of U^{235} is:

$$N_2 = N_o \exp(-\frac{0.693\,t}{T_2})$$

Dividing the above two equations:

$$\frac{N_1}{N_2} = \exp(-\frac{0.693\,t}{T_1} + \frac{0.693\,t}{T_2})$$

or, since:

$$\frac{N_1}{N_2} = \frac{1}{0.072} = 139$$

$$\ln 139 = 4.934 = 0.693\,t(\frac{1}{T_2} - \frac{1}{T_1}) = \frac{0.693(T_1 - T_2)}{T_1 T_2}\,t$$

or:

$$t = \frac{4.934}{0.693}\ \frac{T_1 T_2}{(T_1 - T_2)}$$

Numerically:

$$t = \frac{4.934 \times 4.51 \times 10^9 \times 7.13 \times 10^8}{0.693 \times (45.1 - 7.13) \times 10^8}$$

and finally:

$$\boxed{t = 6 \times 10^9 \text{ years}}$$

Answer

****

(†) Radiological Health Handbook, U.S. Dept. of Health, Education and Welfare, PB 121784R, 1960.

80

Radioactive element A decays into element B which in turn decays into element C. Element A has a decay constant λ_a, B has a decay constant λ_b and C is stable. Initially, the sample contains a mixture of A, B and C in the amounts A_o, B_o, and C_o.

(a) Derive an equation which gives the time at which the amount of element B is maximum.

(b) Using the results of part (a), find an expression for the ratio of the activities of elements A and B for the case when $B_o = 0$, and if the half-life of A is longer than the half-life of B (transient equilibrium). Further, consider the case when not only $T_a > T_b$, but also T_a is extremely long, i.e., $T_a >> T_b$. What would happen then (secular equilibrium)?

****

Let A represent the atoms of element A existing at time t. Then:

$$\frac{dA}{dt} = -\lambda_a A$$

or:

$$A = A_o e^{-\lambda_a t} \qquad (1)$$

If the atoms of element B present at time t are represented by B, then:

$$\frac{dB}{dt} = \lambda_a A - \lambda_b B \qquad (2)$$

or:

$$\frac{dB}{dt} + \lambda_b B = A_o \lambda_a e^{-\lambda_a t}$$

The integrating factor of this differential equation is $e^{\lambda_b t}$:

$$e^{\lambda_b t}\frac{dB}{dt} + \lambda_b B e^{\lambda_b t} = \lambda_a A_o e^{-\lambda_a t}\, e^{\lambda_b t}$$

$$\therefore \qquad \frac{d}{dt}\left(Be^{\lambda_b t}\right) = \lambda_a A_o e^{(\lambda_b - \lambda_a)t}$$

Integrating:

$$Be^{\lambda_b t} = \frac{\lambda_a A_o}{\lambda_b - \lambda_a} e^{(\lambda_b - \lambda_a)t} + \text{constant}$$

At $t = 0$, $B = B_o$, and the constant is:

$$\text{constant} = B_o - \frac{\lambda_a A_o}{\lambda_b - \lambda_a}$$

and:

$$B(t) = \frac{\lambda_a A_o}{\lambda_b - \lambda_a} e^{-\lambda_a t} + B_o e^{-\lambda_b t} - \frac{\lambda_a A_o}{\lambda_b - \lambda_a} e^{-\lambda_b t}$$

which is better expressed:

$$B(t) = \frac{\lambda_a A_o}{\lambda_b - \lambda_a}\left[e^{-\lambda_a t} - e^{-\lambda_b t}\right] + B_o e^{-\lambda_b t} \qquad (3)$$

The maximum of B is obtained by making: $\frac{dB}{dt} = 0$

or:

$$\frac{\lambda_a A_o}{\lambda_b - \lambda_a}\left[-\lambda_a e^{-\lambda_a t} + \lambda_b e^{-\lambda_b t}\right] - B_o \lambda_b e^{-\lambda_b t} = 0$$

Solving for t:

$$t = \frac{1}{\lambda_b - \lambda_a} \ln \left\{ \frac{\lambda_a\lambda_b(A_o + B_o) - B_o\lambda_b^2}{\lambda_a^2 A_o} \right\} \qquad (4)$$

<u>Discussion</u>:

Since logarithms are defined only for positive numbers, one must have in equation (4):

$$\lambda_a\lambda_b(A_o + B_o) - B_o\lambda_b^2 > 0$$

or: $$\lambda_a A_o + \lambda_a B_o > B_o \lambda_b$$

$$\therefore \quad \frac{A_o}{B_o} > \frac{\lambda_b - \lambda_a}{\lambda_a} = \frac{\lambda_b}{\lambda_a} - 1$$

and finally:

$$1 - \frac{\lambda_b}{\lambda_a} > \frac{A_o}{B_o} \qquad (5)$$

But A_o/B_o is essentially positive, and thus:

$$1 - \frac{\lambda_b}{\lambda_a} > 0, \quad \text{or:} \quad \frac{\lambda_b}{\lambda_a} < 1, \quad \text{i.e.:} \quad \boxed{\lambda_b < \lambda_a}$$

This last relation means $T_b > T_a$, indicating that a maximum amount of element B will occur only if the half-life of the daughter is longer than the half-life of the parent. Moreover, equation (4) can be rewritten:

$$\boxed{t = -\frac{1}{\lambda_a - \lambda_b} \ln \left\{ \frac{\lambda_a\lambda_b(A_o + B_o) - B_o\lambda_b^2}{\lambda_a^2 A_o} \right\}} \qquad (6)$$

where now the factor $1/(\lambda_a - \lambda_b)$ is positive.

Three cases can be distinguished:

(i) $\boxed{t > 0}$, the maximum will occur at some future time. For this, the expression within the bracket must be less than 1, or:

$$\lambda_a\lambda_b(A_o + B_o) - B_o\lambda_b^2 < \lambda_a^2 A_o$$

$$\therefore \quad \lambda_a A_o(\lambda_a - \lambda_b) > \lambda_b B_o(\lambda_a - \lambda_b)$$

and finally:

$$\frac{A_o}{B_o} > \frac{\lambda_b}{\lambda_a}$$

Then, when the initial amounts of elements A and B satisfy the above relation, the amount of atoms of element B is increasing and will reach a maximum value for some $t > 0$ given by (6).

(ii) $\boxed{t < 0}$. Conversely, if the quantity within the bracket is positive, the logarithm will be positive, and the value of t given by (6) will be negative, indicating that the maximum of B occurred at some time in the past. Proceeding as above, this condition will occur when the initial amounts of A and B now satisfy the relation:

$$\frac{A_o}{B_o} < \frac{\lambda_b}{\lambda_a}$$

(iii) $\boxed{t = 0}$ Finally, if the initial amounts satisfy the condition:

$$\frac{A_o}{B_o} = \frac{\lambda_b}{\lambda_a}$$

the quantity within the bracket is equal to 1, and equation (6) yields $t = 0$, indicating that the maximum of B occurs precisely at the initial time.†

*

(b) If $T_a > T_b$, then $\lambda_a < \lambda_b$, and the exponential $e^{-\lambda_b t}$ will go to zero faster than $e^{-\lambda_a t}$. Equation (1) remains the same:

$$A(t) = A_o e^{-\lambda_a t}$$

but after some time has elapsed, equation (2) becomes, since $B_o = 0$:

$$B(t) = \frac{\lambda_a A_o}{\lambda_b - \lambda_a} \quad e^{-\lambda_a t}$$

indicating that the number of atoms of element B is decreasing, in agreement with the previous discussion that a maximum could not exist unless $\lambda_a > \lambda_b$. The ratio of the activities is:

$$\frac{R_a}{R_b} = \frac{\lambda_a A}{\lambda_b B} = \frac{\lambda_b - \lambda_a}{\lambda_b} = 1 - \frac{T_b}{T_a}$$

Hence, although A and B are changing with time, the ratio A/B as well as the ratio of the activities is a constant. This condition is called *transient equilibrium*.

In particular, if $T_a >> T_b$, the last term in the above equation can be neglected and:

$$\frac{R_a}{R_b} \cong 1, \qquad \text{or:} \qquad \lambda_a A \cong \lambda_b B$$

indicating that the activities of parent and daughter are the same. Equation (2) becomes then:

$$\frac{dB}{dt} = 0, \qquad \text{or:} \qquad B = \text{constant}$$

i.e., element B is formed at the same rate as it decays, so that its amount remains constant. This condition is called *secular equilibrium*.

****

(†) The above discussion is based on the assumption that $B_o \neq 0$, since for the case $B_o = 0$ the problem is trivial. In fact, equation (4) becomes:

$$t = \frac{1}{\lambda_b - \lambda_a} \ln \left\{ \frac{\lambda_b}{\lambda_a} \right\}$$

which gives a positive value for t regardless of the values of λ_a and λ_b.

81

A sample initially contains 10 milligrams of pure $_{20}Ca^{47}$.
(a) Find its initial activity in curies.
(b) Find the activity in curies of the sample, i.e., parent plus all daughter products, after two days have elapsed.
(c) Find the time at which the amount of the first daughter product reaches a maximum.

****

(a) As found in the literature,† $_{20}Ca^{47}$ decays to $_{21}Sc^{47}$ by beta emission, with a half-life $T_a = 4.8$ days:

$$_{20}Ca^{47} \rightarrow {}_{21}Sc^{47} + \beta^- + \nu$$

Sc^{47} also decays by beta emission into the stable isotope Ti^{47}, with a half-life of 3.43 days:

$$_{21}Sc^{47} \rightarrow {}_{22}Ti^{47} + \beta^- + \nu$$

The initial activity corresponds to the Ca^{47} alone, and is:

$$A = \lambda_a A_o$$

where:

$$\lambda_a = \frac{0.693}{T_a}, \qquad \text{and:} \qquad A_o = \frac{N_o W}{M}$$

Here: N_o = Avogadro's number
W = mass of sample = 10×10^{-3} gm
M = atomic mass of Ca^{47} = 46.9694 amu

Then:

$$A = \frac{0.693}{T_a} \frac{N_o W}{M} = \frac{0.693 \times 6.023 \times 10^{23} \times 10 \times 10^{-3}}{4.8 \times 24 \times 3600 \times 46.9694} = 2.14 \times 10^{14} \frac{\text{dis}}{\text{sec}}$$

and since: 1 curie = 3.7×10^{10} dis/sec (by definition)

finally:

$$A = \frac{2.14 \times 10^{14}}{3.7 \times 10^{10}} = \boxed{5.78 \times 10^3 \text{ curies}}$$ Ans.(a)

(b) After some time has elapsed, the total activity will be the sum of the activities of the parent (Ca^{47}) and the daughter (Sc^{47}):

$$A_t = \lambda_a A(t) + \lambda_b B(t) \qquad (1)$$

where A(t) and B(t) represent, respectively, the number of atoms of parent and daughter existing at time t. From the previous problem, then:

$$A(t) = A_o e^{-\lambda_a t}$$

$$B(t) = \frac{\lambda_a A_o}{\lambda_b - \lambda_a} [e^{-\lambda_a t} - e^{-\lambda_b t}]$$

and (1) becomes:

$$A_t = Ae^{-\lambda_a t} + A \frac{\lambda_b}{\lambda_b - \lambda_a} [e^{-\lambda_a t} - e^{-\lambda_b t}]$$

where $A = \lambda_a A_o$ is the initial activity calculated in part (a).

(†) Radiological Health Handbook, U.S.Dept. of Health, Education and Welfare, PB 121784 R, 1960, pp.238-39.

Numerically: $\lambda_a = \frac{0.693}{4.8} = 0.1445\ d^{-1}$

$$\lambda_b = \frac{0.693}{3.43} = 0.202\ d^{-1}$$

$$\frac{\lambda_b}{\lambda_b - \lambda_a} = \frac{0.202}{0.202 - 0.1445} = \frac{0.202}{0.0575}$$

$$\lambda_a t = 0.1445 \times 2 = 0.289$$

$$\lambda_b t = 0.2020 \times 2 = 0.404$$

Then:

$$A_t = A\ [e^{-0.289} + \frac{0.2020}{0.0575} e^{-0.289} - \frac{0.2020}{0.0575} e^{-0.404}]$$

$$= A\ [0.7485 + \frac{0.2020}{0.0575} \times 0.0815] = 1.0345\ A$$

and finally: $\boxed{A_t = 5970 \text{ curies}}$ Ans.(b)

(c) Referring again to the previous problem, the time at which the daughter product is maximum is given by equation (4), and for $B_o = 0$ we have:

$$t = \frac{1}{\lambda_b - \lambda_a} \ln \{\frac{\lambda_b}{\lambda_a}\}$$

Substituting numerical values:

$$t = \frac{1}{0.0575} \ln [\frac{0.202}{0.1445}] = \frac{\ln 1.4}{0.0575}$$

or: $\boxed{t = 5.85 \text{ days}}$ Ans.(c)

****

82

Radiocarbon dating is based on the assumption that the ratio of C^{12} to C^{14} in the earth's atmosphere has remained constant for thousands of years. Assume that the ratio is 10^6. A sample of a wood relic of unknown age is burned in a reducing flame to obtain pure carbon, and it is found that 1 gram of that carbon has an activity of 3 microcuries. Determine the age of the relic.

****

During the period the plant is alive, the ratio of C^{12} to C^{14} remains constant, but after its death the plant ceases to take in atmospheric carbon dioxide, and the percentage of C^{14} in the wood decreases, as the number of atoms of C^{14} changes following the exponential law of decay:

$$N = N_o e^{-\lambda t}$$

The initial number of atoms of carbon per gram of sample is:

$$N_o = \frac{\mathcal{N}_o}{M}$$

where: $\mathcal{N}_o$ = Avogadro's number, M = atomic mass.

Of these, one in a million is C^{14}; then:

$$N_o = 10^{-6} \frac{\mathcal{N}_o}{M} \qquad (1)$$

Now, the activity of the sample is:

$$A = \lambda N = \lambda N_o e^{-\lambda t}$$

and using (1) and the definition of λ:

$$A = \frac{0.693}{T_{1/2}} \; 10^{-6} \frac{\mathcal{N}_o}{M} e^{-\lambda t}$$

or:

$$e^{\lambda t} = \frac{0.693 \mathcal{N}_o}{MAT_{1/2} \times 10^6}$$

and solving for t:

$$t = \frac{T_{1/2}}{0.693} \ln\left\{ \frac{0.693 \mathcal{N}_o}{MAT_{1/2} \times 10^6} \right\}$$

Numerically:

$$t = \frac{5568}{.693} \ln\left\{ \frac{0.693 \times 6.023 \times 10^{23}}{12 \times 3 \times 10^{-6} \times 3.7 \times 10^{10} \times 5568 \times 3.157 \times 10^7 \times 10^6} \right\}$$

or: $\boxed{t = 4630 \text{ years}}$ <u>Answer</u>

****

83

K^{40} has a half-life of 1.25×10^9 years and a decay scheme as indicated in the figure. Calculate: (a) the disintegration constant in sec^{-1}; (b) the number of beta particles emitted per second by one gram of K^{40}; (c) the number of gammas emitted by one gram of natural potassium per second, knowing that the abundance of K^{40} in nature is 0.0119%.

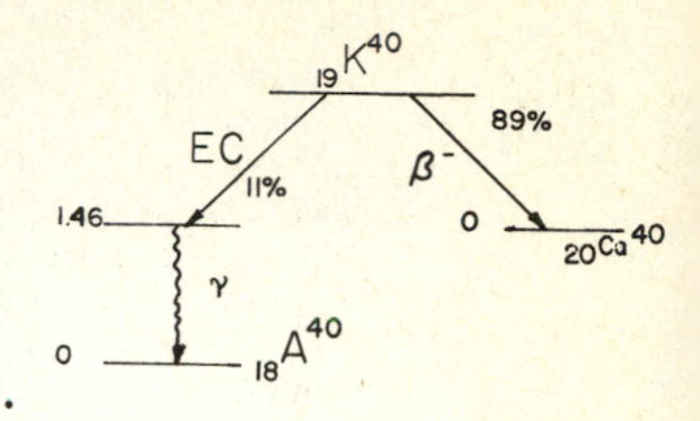

****

(a) The disintegration constant is:

$$\lambda = \frac{\ln 2}{T_{1/2}} = \frac{0.693}{1.25 \times 10^9 (y) \times 3.157 \times 10^7 \text{ (sec/y)}}$$

or:

$$\boxed{\lambda = 1.76 \times 10^{-17} \text{ sec}^{-1}}$$ Ans.(a)

(b) The number of atoms per gram of K^{40} is:

$$N = \frac{N_o}{M} = \frac{6.023 \times 10^{23}}{39.964} = 1.505 \times 10^{22} \text{ atoms/gm}$$

Since beta decay occurs in 89% of the disintegrations, the beta activity is:

$$A_\beta = \lambda N \times 0.89 = 1.76 \times 10^{-17} \times 1.505 \times 10^{22} \times 0.89$$

or:

$$\boxed{A_\beta = 2.36 \times 10^5 \text{ part/sec x gm}}$$ Ans.(b)

(c) For natural potassium, M = 39.1 amu, and the number of atoms per gram is:

$$N = \frac{N_o}{M} = \frac{6.023 \times 10^{23}}{39.1} = 1.54 \times 10^{22} \text{ atom/gm}$$

But only 0.0119% of the atoms of natural potassium are K^{40}, and moreover, only 11% of the disintegrations will result in the emission of a gamma. Thus, the gamma activity is:

$$A_\gamma = \lambda N \times 0.11 \times 0.000119$$

$$= 1.76 \times 10^{-17} \times 1.54 \times 10^{22} \times 0.11 \times 0.000119$$

Or:

$$\boxed{A_\gamma = 3.55 \text{ gammas/sec x gm}}$$ Ans.(c)

****

84

A mineral contains the following mass percentages of radioactive elements: U^{238}, 19.5%; Th^{232}, 58%. It also contains, among other substances, 1.3% of lead as a mixture of isotopes 206 and 208.

(a) Assuming the lead to be entirely of radioactive origin, determine the age of the mineral.

(b) The decay of the radioactive elements into lead involves the emission of alpha particles. The helium thus formed is usually occluded in the mineral, and can be recovered and its volume measured. Determine the mass in grams, and the volume at STP, of the He contained in this mineral per gram of rock.

****

(a) U^{238} is the first element of the (4n+2) radioactive series, which terminates in the stable isotope Pb^{206}. The half-life of the parent is extremely long, $T_a = 4.51 \times 10^9$ years, as compared with the half-lives of the daughter elements, and hence the system is under conditions of secular equilibrium. Then, the activities of all members of the chain are the same, and consequently the atoms of Pb^{206} are formed at the same rate as the atoms of U^{238} decay.

Let: Pb^{206} = number of atoms of Pb^{206}

λ_u = disintegration constant of U^{238}

U = number of atoms of U^{238}

W_u = mass of U^{238} in the sample

M_u = atomic mass of $U^{238} \approx 238$

The rate of formation of Pb^{206} is:

$$\frac{dPb^{206}}{dt} = \lambda_u U$$

Moreover, under secular equilibrium conditions it can be assumed that the amount of uranium does not change appreciably, i.e., $\lambda_u U$ is constant. Integration of above equation yields:

$$Pb^{206} = \lambda_u U \Delta t$$

where Δt is the desired age of the mineral.

If N_o is Avogadro's number, M_6 the atomic mass of Pb^{206}, and W_6 the mass of Pb^{206} in the sample, then:

$$U = \frac{N_o W_u}{M_u}, \qquad Pb^{206} = \frac{N_o W_6}{M_6}$$

and above equation becomes:

$$\frac{N_o M_6}{M_6} = \lambda_u \frac{N_o M_u}{M_u} \Delta t$$

Solving for Δt:

$$\boxed{\Delta t = \frac{W_6}{W_u}\frac{M_u}{M_6}\frac{1}{\lambda_u}} \qquad (1)$$

The ratio of the masses can be replaced by the ratio of the respective mass percentages, while the atomic masses can be approximated by the respective mass numbers. If a represents the fraction of Pb ^{206} in the sample, by mass, then:

$$\Delta t = \frac{a}{0.195}\;\frac{238}{206}\;\frac{4.51 \times 10^9}{0.693}$$

$$\therefore \qquad \Delta t = 3.85 \times 10^{10}\ a \qquad (2)$$

Similarly, Th^{232} is the first element of the 4n radioactive series, which terminates in the stable isotope Pb^{208}. The half-life of Th^{232} is $T_b = 1.39 \times 10^{10}$ y. Under the assumptions of secular equilibrium, an expression similar to (1) is obtained:

$$\Delta t = \frac{W_8}{W_t} \frac{M_t}{M_8} \frac{1}{\lambda_t}$$

or, numerically:

$$\Delta t = \frac{b}{0.58} \; \frac{232}{208} \; \frac{1.39 \times 10^{10}}{0.693}$$

$$\Delta t = 3.85 \times 10^{10}\ b \qquad (3)$$

where b is the fraction by mass of Pb^{208} in the sample. But the age of the rock is only one, so that (2) and (3) must be equal. Thus: a = b, and since we know that a + b = 0.013:

$$a = b = 0.0065 = 6.5 \times 10^{-3}$$

Notice that U^{235} must also be present in the mineral, since natural uranium contains 0.72% of U^{235}. This small amount can be ignored, and moreover, the end product of the disintegration of U^{235} is Pb^{207} which is not considered here.

The age of the rock can now be calculated using either (2) or (3):

$$\Delta t = 3.85 \times 10^{10} \times 6.5 \times 10^{-3}$$

or:

$$\boxed{\Delta t = 2.5 \times 10^{8} \text{ years}} \qquad \underline{\text{Ans.(a)}}$$

(b) Consider first the disintegration of U^{238}. As shown by the difference in the mass numbers, the decay processes from U^{238} to Pb^{206} imply the emission of 8 alpha particles:

$$U^{238} \rightarrow Pb^{206} + 8He^4$$

or, in other words, each atom of Pb^{206} present indicates the presence of 8 atoms of He^4. But each atom of Pb^{206} has a mass $206/N_o$, while each atom of He^4 has a mass $4/N_o$. Thus:

$$\frac{206}{N_o} \text{ gm of } Pb^{206} \quad \text{imply:} \quad \frac{4}{N_o} \times 8 \text{ gm of } He^4$$

$$\text{or:} \qquad 1 \text{ gm of } Pb^{206} \rightarrow \frac{32}{206} \text{ gm of } He^4$$

and finally, since the fraction of Pb^{206} in the sample is a, the mass of the He per gram of sample due to the disintegration of U^{238} is:

$$He(U) = \frac{32}{206}\ a \qquad (4)$$

Likewise, the decay of Th^{232} into Pb^{208} implies the ejection of 6 alpha particles:

$$Th^{232} \rightarrow Pb^{208} + 6\ He^4$$

and using the same reasoning the amount of He per gram of sample due to the decay of Th^{232} is:

$$He(Th) = \frac{24}{208}\ b \qquad (5)$$

The total amount of He in the rock, per unit mass of mineral, is obtained adding (4) and (5):

$$He = \frac{32}{206}\,a + \frac{24}{208}\,b$$

In this case, $a = b = 6.5 \times 10^{-3}$, and:

$$He = \left[\frac{32}{206} + \frac{24}{208}\right] \times 6.5 \times 10^{-3}$$

or:

$$\boxed{He = 1.76 \times 10^{-3} \text{ gm/gm of sample}} \qquad \underline{\text{Ans.(b)}}$$

Finally, at S.T.P. the gram molecular weight of any gas occupies a volume of 22414 cm^3. Hence:

$$V_{He} = \frac{He}{4} \times 22414 \quad cm^3 \text{ /gm of sample}$$

or:

$$V_{He} = \frac{1.76 \times 10^{-3} \times 22.414 \times 10^3}{4}$$

$$\therefore \qquad \boxed{V_{He} = 9.86 \text{ cm}^3\text{/gm of sample}} \qquad \underline{\text{Ans.(b)}}$$

It is interesting to verify the assumption that the change in the amount of U^{238} during the time Δt is small. We have:

$$\lambda_u t = \frac{0.693 \times 2.5 \times 10^8}{4.51 \times 10^9} = 0.0384$$

Using the law of radioactive decay:

$$N = N_o e^{-\lambda_u t}$$

one has:

$$\frac{N}{N_o} = \frac{W_u(t)}{W_u(0)} = e^{-\lambda_u t} \cong 1 - \lambda_u t = 1 - 0.0384 = 0.9616$$

or:

$$W_u(t) = 0.9616\ W_u(0)$$

which justifies the assumption.

****

6

NUCLEAR PHYSICS

(b) Nuclear Reactions

85

Show that in a nuclear reaction where the product particle is ejected at an angle of 90° with the direction of the bombarding particle, the Q-value is expressed:

$$Q = K_y\left(1 + \frac{m_y}{M_Y}\right) - K_x\left(1 - \frac{m_x}{M_X}\right)$$

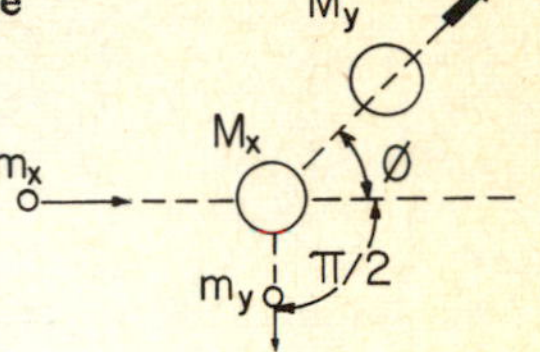

****

By conservation of energy: $K_x + Q = K_y + K_Y$ (1)

By conservation of momentum:

vertical component: $M_Y v_Y \sin \phi = m_y v_y$

horizontal component: $M_Y v_Y \cos \phi = m_x v_x$

Squaring and adding: $M_Y^2 v_Y^2(\sin^2\phi + \cos^2\phi) = m_y^2 v_y^2 + m_x^2 v_x^2$

or, since the particles are not relativistic:

$$M_Y \frac{1}{2} M_Y v_Y^2 = \frac{1}{2} m_y v_y^2 m_y + \frac{1}{2} m_x v_x^2 m_x$$

$$M_Y K_Y = m_y K_y + m_x K_x \qquad (2)$$

Eliminating K_Y between (1) and (2):

$$K_x + Q = \frac{m_y}{M_Y} K_y + \frac{m_x}{M_Y} K_x + K_y$$

or: $$\boxed{Q = K_y\left(1 + \frac{m_y}{M_Y}\right) - K_x\left(1 - \frac{m_x}{M_Y}\right)}$$ Q.E.D.

****

86

Calculate the binding energy of the last neutron in O^{16} and in C^{13}. Use the following atomic masses, based on C^{12}:

$${}_0n^1 = 1.008665 \text{ amu}, \quad {}_6C^{13} = 13.003354 \text{ amu}, \quad {}_8O^{15} = 15.003070 \text{ amu}$$

$${}_8O^{16} = 15.994915 \text{ amu}.$$

****

By definition of binding energy:

$$BE = (\text{mass of neutron} + \text{mass } O^{15}) - (\text{mass } O^{16})$$

$$= (1.008665 + 15.003070) - 15.994915 = 0.01682 \text{ amu}$$

and using the conversion factor: 1 amu = 931.5 Mev

$$\boxed{BE = 15.68 \text{ Mev}}$$ Ans.

Likewise, for C^{13}:

$$BE = (\text{mass of neutron} + \text{mass of } C^{12}) - (\text{mass of } C^{13})$$

$$= (1.008665 + 12.000000) - 13.003354 = 0.005311 \text{ amu}$$

or: $$\boxed{BE = 4.95 \text{ Mev}}$$ Ans.

Consequently, it will be much more difficult to separate a neutron from an O^{16} nucleus than from a C^{13} nucleus. This was to be expected, since O^{16} is a "doubly magic" nucleus.

****

87 Find the total binding energy of $_2He^4$ and $_2He^5$. Which one of these nucleii is more stable ?

****

The binding energy is the equivalent to the mass difference of the separate nucleons forming the nucleus and the nuclear mass:

$$\frac{BE}{c^2} = m = Zm_p + (A-Z)m_n - M$$

However, the masses that are tabulated are the atomic masses and not the nuclear masses. Consequently, by adding and subtracting Z electron masses to the right hand side of the above equation, one has:

$$\boxed{\frac{BE}{c^2} = Z\,{}_1M^1 + (A-Z)m_n - {}_ZM^A} \quad \text{amu} \qquad (1)$$

where: $_1M^1$ = mass of atom of mass number 1 and atomic number 1, i.e., hydrogen

$_ZM^A$ = atomic mass of atom of mass number A and atomic number Z.

Using atomic masses based on oxygen 16[†], one has:

atomic mass of	$_2He^4$	= 4.003873	amu
" " "	$_2He^5$	= 5.013888	"
" " "	$_1H^1$	= 1.008145	"
mass of neutron	m_n	= 1.008986	"

For He^4 equation (1) yields:

$$\frac{BE}{c^2} = 2 \times 1.008145 + 2 \times 1.008986 - 4.003873 = 0.030389 \text{ amu}$$

or: $BE = 0.030389 \times 931.5 = \boxed{28.3 \text{ Mev}}$ Answer

For He^5, similarly:

$$\frac{BE}{c^2} = 2 \times 1.008145 + 3 \times 1.008986 - 5.013888 = 0.02936 \text{ amu}$$

or: $\boxed{BE = 27.4 \text{ Mev}}$ Answer

The nucleus of He^4 (alpha particle) is then more stable, since more energy is required to break it into its components.

****

(†) See for example, Radiological Health Handbook, U.S. Dept. of Health, Education and Welfare, PB 121784R, 1960, p.79.

88 Determine whether the nuclear reaction:

$$_{3}Li^{6} + {}_{0}n^{1} \rightarrow ({}_{3}Li^{7}) \rightarrow {}_{1}H^{3} + {}_{2}He^{4}$$

can be caused by slow neutrons.

****

The Q-value of the nuclear reaction is given by the difference between the initial and final masses. Using atomic masses based on C^{12} one has:

$$\text{initial masses:} \quad {}_{3}Li^{6} = 6.015125, \quad {}_{0}n^{1} = 1.008665, \quad \text{sum} = 7.023790$$

$$\text{final masses:} \quad {}_{2}He^{4} = 4.002603, \quad {}_{1}H^{3} = 3.016050, \quad \text{sum} = 7.018653$$

Thus: $Q = 7.023790 - 7.018653 = 0.005137$ amu

or: $Q = 0.005137 \times 931.5 = 4.77$ Mev

Since $Q > 0$, the reaction is exothermic and will occur even if the neutrons have zero kinetic energy.

****

89

Complete the following nuclear reactions:

$$_3Li^6 + {}_1p^1 \rightarrow (?) \rightarrow {}_2He^3 + ?$$

$$_{18}Ar^{40} + {}_2He^4 \rightarrow (?) \rightarrow ? + {}_0n^1$$

$$_9F^{19} + ? \rightarrow (?) \rightarrow {}_8O^{16} + {}_2He^4$$

$$_ZX^A(d,2n) ?$$

****

Considering that both charge and number of nucleons must be conserved, the nuclear reactions are written at once as follows:

$$_3Li^6 + {}_1p^1 \rightarrow ({}_4Be^7) \rightarrow {}_2He^3 + {}_2He^4$$

$$_{18}Ar^{40} + {}_2He^4 \rightarrow ({}_{20}Ca^{44}) \rightarrow {}_{20}Ca^{43} + {}_0n^1$$

$$_9F^{19} + {}_1H^1 \rightarrow ({}_{10}Ne^{20}) \rightarrow {}_8O^{16} + {}_2He^4$$

$$_ZX^A(d, 2n)\ {}_{Z+1}X^A$$

****

90 Write ten nuclear reactions which you might expect to take place if the target is ${}_{13}Al^{27}$. Include the compound nucleus.

****

For low energy nuclear reactions, i.e., for energies of the order of 20 Mev or less, the bombarding particle can be any of the following: proton, neutron, deuteron, alpha particle, or gamma photon. The reactions are easily completed by taking into account that both charge (i.e., Z) and number of nucleons (i.e. A) must be conserved. Then:

$${}_{13}Al^{27} + {}_{1}H^{1} \rightarrow ({}_{14}Si^{28}) \rightarrow {}_{14}Si^{28} + \gamma$$

$${}_{13}Al^{27} + {}_{1}H^{1} \rightarrow ({}_{14}Si^{28}) \rightarrow {}_{14}Si^{27} + {}_{0}n^{1}$$

$${}_{13}Al^{27} + {}_{0}n^{1} \rightarrow ({}_{13}Al^{28}) \rightarrow {}_{12}Mg^{27} + {}_{1}H^{1}$$

$${}_{13}Al^{27} + {}_{0}n^{1} \rightarrow ({}_{13}Al^{28}) \rightarrow {}_{13}Al^{28} + \gamma$$

$${}_{13}Al^{27} + {}_{1}H^{2} \rightarrow ({}_{14}Si^{29}) \rightarrow {}_{13}Al^{28} + {}_{1}H^{1}$$

$${}_{13}Al^{27} + {}_{1}H^{2} \rightarrow ({}_{14}Si^{29}) \rightarrow {}_{12}Mg^{25} + {}_{2}He^{4}$$

$${}_{13}Al^{27} + {}_{1}H^{2} \rightarrow ({}_{14}Si^{29}) \rightarrow {}_{14}Si^{28} + {}_{0}n^{1}$$

$${}_{13}Al^{27} + {}_{2}He^{4} \rightarrow ({}_{15}P^{31}) \rightarrow {}_{15}P^{30} + {}_{0}n^{1}$$

$${}_{13}Al^{27} + {}_{2}He^{4} \rightarrow ({}_{15}P^{31}) \rightarrow {}_{14}Si^{30} + {}_{1}H^{1}$$

$${}_{13}Al^{27} + \gamma \rightarrow ({}_{13}Al^{27}) \rightarrow {}_{13}Al^{26} + {}_{0}n^{1}$$

****

91

(a) Show that the threshold kinetic energy of the bombarding particle for the nuclear reaction A(a,b)B is given by:

$$K = -Q\left[\frac{m_a + m_b + m_A + m_B}{2m_A}\right]$$

(b) Calculate the Q-value and the threshold energy for the reaction: $Li^7(\alpha,n)B^{10}$. Use atomic masses based on C^{12}.

****

(a) One uses the invariance of the momentum-energy 4-vector:

$$(\Sigma E_i)^2 - (\Sigma p_i c)^2 = \text{constant} \qquad (1)$$

where: ΣE_i = sum of the total energy for all particles in system

$\Sigma \underline{p}_i$ = total momentum of particles of the system

In the LCS, before the collision, the target nucleus is at rest and has total energy:

$$E_A = m_a c^2$$

and momentum $p_A = 0$, while the bombarding particle has total energy E_a and momentum p_a. At the threshold energy, the products will be created at rest in the CMCS, i.e., their total energies will be just their rest masses. Equation (1) is then:

$$(E_a + m_A c^2)^2 - p_a^2 c^2 = (m_b c^2 + m_B c^2)^2$$

or:

$$E_a^2 + (m_A c^2)^2 + 2E_a\, m_A c^2 - p_a^2 c^2 = (m_b c^2 + m_B c^2)^2$$

But: $E_a^2 - p_a^2 c^2 = (m_a c^2)^2$

and: $E_a = m_a c^2 + K$

then:

$$(m_a c^2)^2 + (m_A c^2)^2 + 2\, m_a c^2\, m_A c^2 + 2K\, m_A c^2 = (m_b c^2 + m_B c^2)^2$$

$$\therefore \quad (m_a c^2 + m_A c^2)^2 + 2K\, m_A c^2 = (m_b c^2 + m_B c^2)^2$$

Solving for K:

$$K = \frac{(m_b c^2 + m_B c^2)^2 - (m_a c^2 + m_A c^2)^2}{2\, m_A c^2}$$

which is also written:

$$K = \frac{(m_b c^2 + m_B c^2 + m_a c^2 + m_A c^2)(m_b c^2 + m_B c^2 - m_a c^2 - m_A c^2)}{2 m_A c^2}$$

But the Q-value of the reaction is the difference between initial and final masses, so that the second factor in the numerator is -Q. Then:

$$\boxed{K = -Q\left[\frac{m_a + m_b + m_A + m_B}{2m_A}\right]} \qquad \text{Q.E.D.}$$

Notice that this formula gives the exact relativistic expression for the threshold energy. Using again the definition of Q-value, this is also written:

$$K = -Q\left[\frac{2(m_a + m_A) - (Q/c^2)}{2m_A}\right]$$

where the term (Q/c^2) is small compared to $(m_a + m_A)$ and can be neglected. Thus, the formula:

$$\boxed{K = -Q\left[\frac{m_a + m_A}{m_A}\right]}$$

which appears very often in the literature.

(b) The atomic masses based on C^{12} are as follows:†

$${}_3Li^7 = 7.01601 \text{ amu} \qquad {}_5B^{10} = 10.01294 \text{ amu}$$

$${}_2He^4 = 4.00260 \text{ amu} \qquad {}_0n^1 = 1.008665 \text{ amu}$$

initial masses: 11.01861 amu , final masses: 11.021605 amu

Then: $Q = 11.01861 - 11.021605 = -0.002995 \text{ amu} = -0.002995 \times 931.5 \text{ Mev}$

or: $\boxed{Q = -2.79 \text{ Mev}}$ <u>Answer (b)</u>

The threshold energy is then:

$$K = +2.79 \frac{22.0402}{14.032} = 4.39 \text{ Mev}$$

or: $\boxed{K = 4.39 \text{ Mev}}$ <u>Answer (b)</u>

****

(†)Handbook of Chemistry and Physics, The Chemical Rubber Co, 1966 (47th edition), Section B.

92 Determine the Q-value and the threshold energy for the nuclear reaction:

$$_{9}F^{19} + n \rightarrow p + {_{8}O^{19}}$$

****

Atomic masses based on C^{12} will be used:

$_{9}F^{19}$ = 18.998405 amu , $_{8}O^{19}$ = 19.003578 amu, n = 1.0086654 amu,

p = 1.0072766 amu

The Q-value is:

Q = atomic masses of initial particles - atomic masses of final particles.

Thus: $Q = (18.998405 + 1.008665) - (1.007276 + 19.003578)$

$= - 0.003784 \text{ amu} = - 0.003784 \times 931.5 \text{ Mev}$

or: $\boxed{Q = - 3.52 \text{ Mev}}$ <u>Answer</u>

The reaction being endoergic, the bombarding particle should provide not only for the mass difference but also for the kinetic energy of the reaction products, which have to be ejected if momentum is to be conserved. The threshold energy is then:

$$E_{th} = |Q|(1 + \frac{m_x}{M_X}) = 3.52\ (1 + \frac{1}{19})$$

or: $\boxed{E_{th} = 3.71 \text{ Mev}}$ <u>Answer</u>

****

93

In the nuclear reaction: $C^{13} + p \rightarrow (N^{14})^*$, protons of 1.75 Mev are needed to produce N^{14} in an excited state.

(a) Calculate the energy, K_{cm}, carried by the center of mass.

(b) Calculate the excitation energy, E_e, i.e., the energy of the compoun[d] nucleus above the ground state.

(c) $(N^{14})^*$ decays to the ground state by gamma emission. Calculate the energy of the recoil nucleus and the energy of the gamma photon.

****

(a) Let m_x be the mass of the bombarding proton, M_X the mass of the target atom of C initially at rest. The velocity of the center of mass is:

$$v_{cm} = \frac{m_x v_x}{m_x + M_X}$$

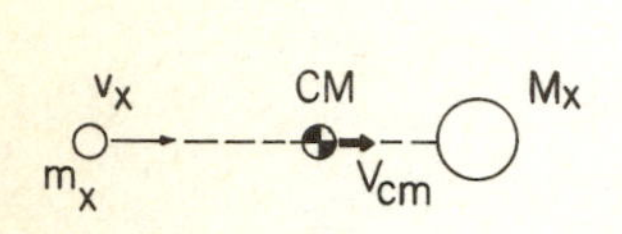

The kinetic energy carried by the center of mass is then:

$$K_{cm} = \frac{1}{2}(m_x + M_X)v_{cm}^2 = \frac{1}{2}\frac{m_x^2 v_x^2}{(m_x + M_X)}$$

or:

$$K_{cm} = K_x \frac{m_x}{m_x + M_X} \qquad (1)$$

Using atomic masses based on oxygen 16:

$$M_X = 13.0074900$$
$$m_x = 1.0081451$$

equation (1) yields:

$$K_{cm} = 1.75 \times \frac{1.0081}{14.0156} = \boxed{0.126 \text{ Mev}} \qquad \text{Ans.(a)}$$

(b) The energy available in the CMCS before the reaction is the kinetic energy of the bombarding particle minus the kinetic energy of the center of mass computed in part (a):

$$E_a = K_x - K_{cm} = 1.75 - 0.126 = 1.624 \text{ Mev}$$

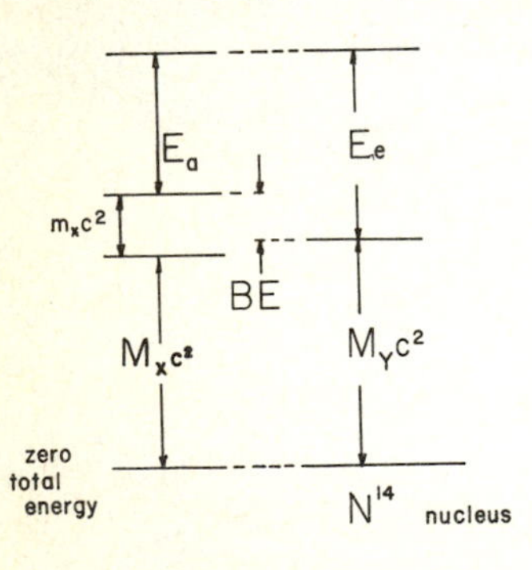

The excitation energy of the N^{14} nucleus will be the sum of this energy and the energy resulting of the difference in mass between the initial and final particles, i.e., the binding energy of the proton:

$$BE = (m_x + M_X) - M_Y$$

$$= (1.0081451 + 13.0074900) - 14.00752[...]$$

$$= 0.0081094 \text{ amu} = 0.0081094 \times 931.5 \text{ M[ev]}$$

or: $\quad BE = 7.54$ Mev

Hence:

$$E_e = E_a + BE = 1.624 + 7.54 = 9.164 \text{ Mev}$$

or: $\boxed{E_e = 9.164 \text{ Mev}}$ $\qquad$ Ans.(b)

(c) When the nucleus decays to the ground state, the gamma photon and the recoil nucleus will move in opposite directions, and will share the energy calculated in part (b). By conservation of energy:

$$E_e = h\nu + E_r \qquad (2)$$

By conservation of momentum:

$$\frac{h\nu}{c} = p, \qquad \text{or:} \qquad h\nu = pc \qquad (3)$$

Eliminating $h\nu$ between (2) and (3), and using also $E_r = p^2/2M$, one gets:

$$\frac{p^2c^2}{2Mc^2} + pc - E_e = 0$$

which has solutions:

$$pc = \frac{-1 \pm \sqrt{1 + \frac{4E_e}{2Mc^2}}}{\frac{2}{2Mc^2}} = Mc^2 \left[-1 + \sqrt{1 + \frac{4E_e}{2Mc^2}}\right]$$

where the minus sign has been dropped since it has no physical meaning, since $pc > 0$.

But: $\quad 4E_e \ll 2Mc^2, \quad$ or: $\quad \frac{4E_e}{2Mc^2} \ll 1$

and the approximation: $\quad (1+x)^{1/2} \simeq 1 + \frac{1}{2}x \quad$ can be used. Then:

$$pc = Mc^2 \left[-1 + 1 + \frac{1}{2}\frac{4E_e}{2Mc^2}\right] = E_e$$

which is equivalent to assume that practically the whole energy is carried away by the photon. The energy of the recoil nucleus is:

$$E_r = \frac{p^2}{2M} = \frac{p^2c^2}{2Mc^2} = \frac{E_e^2}{2Mc^2} = \frac{(9.164)^2}{2 \times 14.0075 \times 931.5} = 3.22 \times 10^{-3} \text{ Mev}$$

or: $\boxed{E_r = 0.0032 \text{ Mev}}$ Ans.(c)

The energy of the gamma photon is, therefore:

$h\nu = E_\gamma = 9.164 - 0.003 \qquad$ or: $\boxed{E_\gamma = 9.161 \text{ Mev}}$ Ans.(c)

****

94

Consider a nuclear reaction between a bombarding particle of mass m_x and a target nucleus of mass M_X at rest. The product nucleus has mass M_Y, and a particle of mass m_y is ejected. None of the particles is relativistic. Let v_{cm} and v'_{cm} be the velocities of the center of mass before and after the collision, respectively. Show that:

$$v'_{cm} = v_{cm}\left[1 + \frac{Q}{(m_y + M_Y)c^2}\right]$$

****

BEFORE

AFTER

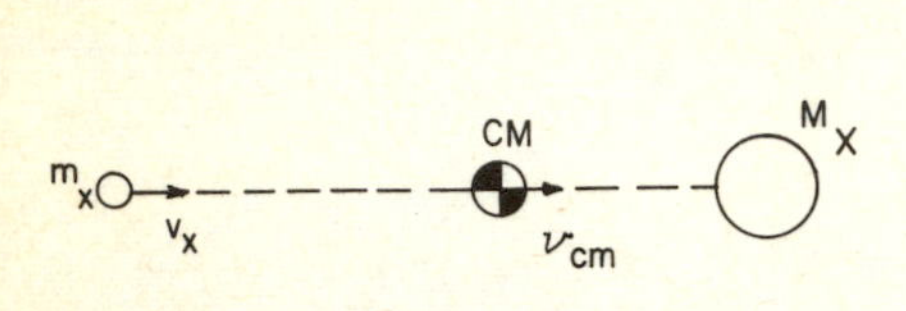

M_Y

CM

v'_{cm}

m_y

By definition of center of mass, its horizontal velocity after the collision is:

$$v'_{cm} = \frac{m_y v_y + M_Y v_Y}{m_y + M_Y} \qquad (1)$$

where v_y, v_Y represent the horizontal components of the velocities of of y and Y. Thus, by conservation of linear momentum in the horizontal direction:

$$m_y v_y + M_Y v_Y = m_x v_x$$

and (1) is written:

$$v'_{cm} = \frac{m_x v_x}{m_y + M_Y} = \frac{m_x v_x}{m_x + M_X}\,\frac{m_x + M_X}{m_y + M_Y}$$

where the first factor is the velocity of the center of mass before the collision. Hence:

$$v'_{cm} = v_{cm}\,\frac{m_x + M_X}{m_y + M_Y} = v_{cm}\,\frac{m_y + M_Y - (m_y + M_Y) + m_x + M_X}{m_y + M_Y}$$

or:

$$v'_{cm} = v_{cm}\left[1 + \frac{(m_x + M_X) - (m_y + M_Y)}{m_y + M_Y}\right]$$

But the Q-value is:

$$\frac{Q}{c^2} = (\text{initial masses}) - (\text{final masses})$$
$$= (m_x + M_X) - (m_y + M_Y)$$

and replacing above:

$$\boxed{v'_{cm} = v_{cm}\left[1 + \frac{Q}{(m_y + M_Y)c^2}\right]} \qquad \text{Q.E.D.}$$

This can be also expressed:

$$v'_{cm} = \frac{v_{cm}}{1 - \dfrac{Q}{(m_x + M_X)c^2}}$$

In practice, the difference between v_{cm} and v'_{cm} is negligible.

****

95

In a nuclear reaction the bombarding particle has rest mass m and kinetic energy K, while the target nucleus has mass M and is at rest. Show that the excitation energy of the compound nucleus is given by:

$$E_e = K\left(\frac{M}{m+M}\right) + BE$$

where BE is the binding energy of the bombarding particle in the compound nucleus.

****

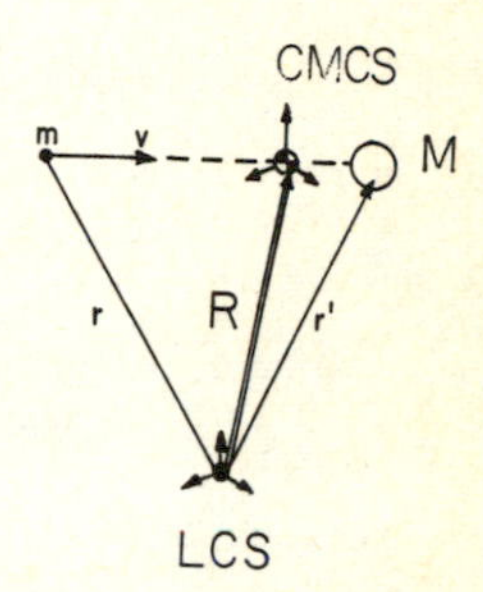

All particles are heavy enough so that relativistic effects can be neglected. The center of mass, CM, of the system m, M, is defined by a vector $\underline{R}$ such that:

$$m\underline{r} + M\underline{r}' = (m+M)\underline{R}$$

Taking time derivatives, since M is at rest:

$$m\dot{\underline{r}} = (m+M)\dot{\underline{R}}$$

or, in the LCS the velocity of the center of mass is:

$$\dot{R} = v_c = \frac{mv}{m+M}$$

Hence, in the CMCS the velocities of the bombarding particle and the target nucleus are, respectively:

$$v_1 = v - v_c = \frac{Mv}{m+M}$$

$$v_2 = - v_c = - \frac{mv}{m+M}$$

Now, in the CMCS, the compound nucleus is at rest and has energy E_e above the ground state. Conservation of energy before and after the formation of the compound nucleus is expressed:

$$\frac{1}{2}\, m\left(\frac{Mv}{m+M}\right)^2 + mc^2 + \frac{1}{2}\, M\left(\frac{mv}{m+M}\right)^2 + Mc^2 = M_c c^2 + E_e$$

where M_c is the mass of the compound nucleus. Therefore:

$$\frac{1}{2}\;\frac{mM}{m+M}\, v^2 + (m+M)c^2 - M_c c^2 = E_e \qquad (1)$$

But the binding energy of the bombarding particle is, by definition:

$$BE = (m+M)c^2 - M_c c^2$$

and:

$$\frac{1}{2}\, mv^2 = \text{kinetic energy of bombarding particle in LCS} = K$$

Then, equation (1) becomes:

$$\boxed{E_e = K\left(\frac{M}{m+M}\right) + BE} \qquad \text{Q.E.D.}$$

****

96

The compound nucleus resulting from bombarding $_3Li^7$ with 0.441-Mev protons decays to its ground state by emission of a very energetic gamma photon.
(a) Write the nuclear reaction.
(b) Calculate the excitation energy of the compound nucleus.
(c) Determine the wavelength of the emitted radiation.

****

(a) Taking into account that both charge and number of nucleons must be conserved, the nuclear reaction is written at once:

$$_3Li^7 + {_1H^1} \rightarrow (_4Be^8) \rightarrow {_4Be^8} + h\nu$$

The $_4Be^8$ nucleus is not stable, and decays almost at once into two alpha particles.

(b) The binding energy of the proton is calculated using atomic masses based on C^{12}:

$$_3Li^7 = 7.016010 \text{ amu}$$
$$_1H^1 = 1.007825 \text{ "}$$

initial masses: 8.023835 amu

final mass, $_4Be^8$ = 8.005310 amu

Subtracting: BE = 0.018525 amu = 0.018525 x 931.5 = 17.25 Mev

Now, the excitation energy is given by:

$$E_e = K\left(\frac{M}{m+M}\right) + BE$$

where K, the kinetic energy of the bombarding particle, is 0.441 Mev. Then:

$$E_e = 0.441\left(\frac{7}{1+7}\right) + 17.25 = 0.386 + 17.25$$

or: $$\boxed{E_e = 17.636 \text{ Mev}}$$ Ans.(b)

(c) Since the recoil energy of the nucleus can be neglected, the photon will carry the excitation energy and:

$$E_e = h\nu = \frac{hc}{\lambda}$$

and from here:

$$\lambda = \frac{hc}{E_e} = \frac{6.626 \times 10^{-27}(\text{erg-sec}) \times 3 \times 10^{10}(\text{cm/sec})}{17.636\ (\text{Mev}) \times 1.6 \times 10^{-6}(\text{erg/Mev}) \times 10^{-8}\ (\text{cm/Å})}$$

or: $$\boxed{\lambda = 7.03 \times 10^{-4}\ \text{Å}}$$ Ans.(c)

****

97 Show that when a nucleus of rest mass M absorbs a photon $h\nu$, the excitation energy of the nucleus is given by:

$$E_e = Mc^2\left(\sqrt{1 + \frac{2h\nu}{Mc^2}} - 1\right)$$

****

The relativistic center of mass of the system formed by the photon and the nucleus is moving with velocity:†

$$\underline{V} = \frac{c^2}{E}\underline{P}$$

where: $\underline{P}$ = total momentum of the particles in the laboratory coordinate system, LCS.
$\underline{E}$ = total energy of the particles in the LCS.

Since the nucleus is at rest: $P = \frac{h\nu}{c}$

and: $E = Mc^2 + h\nu$

$$\text{Thus:} \quad V = \frac{h\nu}{Mc^2 + h\nu}\,c \quad \text{, or:} \quad \beta = \frac{h\nu}{Mc^2 + h\nu} \qquad (1)$$

The center of mass is moving in the same direction as the incident photon, and away from the photon source. Hence, in the center of mass coordinate system, CMCS, the frequency of the photon is changed by the Doppler effect and:

$$\nu' = \sqrt{\frac{1-\beta}{1+\beta}}\;\nu$$

By conservation of energy in the CMCS:

$$Mc^2 + Mc^2(\gamma-1) + h\nu' = Mc^2 + E_e$$

or:

$$E_e = Mc^2(\gamma-1) + h\nu\sqrt{\frac{1-\beta}{1+\beta}}$$

But:

$$\frac{1-\beta}{1+\beta} = \frac{(1-\beta)^2}{1-\beta^2} = \gamma^2(1-\beta)^2$$

and:

$$E_e = Mc^2(\gamma-1) + h\nu(1-\beta)\gamma$$

and since:

$$1 - \beta = 1 - \frac{h\nu}{Mc^2 + h\nu} = \frac{Mc^2}{Mc^2 + h\nu}$$

$$E_e = Mc^2(\gamma-1) + \gamma\,\frac{h\nu\, Mc^2}{Mc^2 + h\nu} = Mc^2\left\{\gamma\left(1 + \frac{h\nu}{Mc^2 + h\nu}\right) - 1\right\}$$

or:

$$E_e = Mc^2\left\{\gamma\,\frac{Mc^2 + 2h\nu}{Mc^2 + h\nu} - 1\right\} \qquad (2)$$

Using (1), an expression for γ is now obtained:

$$\gamma = \frac{1}{\sqrt{1-\beta^2}} = \frac{1}{\sqrt{1 - \dfrac{(h\nu)^2}{(Mc^2 + h\nu)^2}}} = \frac{Mc^2 + h\nu}{\sqrt{(Mc^2)^2 + 2h\nu\, Mc^2}}$$

and substituting in (2):

(†) See, for example, Katz, Introd. to Special Theory of Relativity, Van Nostrand Momentum Book #9, 1964, p.78.

$$E_e = Mc^2 \left\{ \frac{Mc^2 + h\nu}{\sqrt{(Mc^2)^2 + 2h\nu\, Mc^2}} \; \frac{Mc^2 + 2h\nu}{Mc^2 + h\nu} - 1 \right\}$$

$$= Mc^2 \left\{ \frac{(Mc^2 + 2h\nu)\sqrt{(Mc^2)^2 + 2h\nu\, Mc^2}}{Mc^2\,[Mc^2 + 2h\nu]} - 1 \right\}$$

or:

$$\boxed{E_e = Mc^2 \left(\sqrt{1 + \frac{2h\nu}{Mc^2}} - 1\right)} \qquad \text{Q.E.D.}$$

The problem is much more easily solved by using the invariance of the square of the length of the momentum-energy 4-vector:

$$(\Sigma E_i)^2 - (\Sigma p_i c)^2 = \text{constant} \tag{3}$$

where: ΣE_i = total energy of particles of the system

Σp_i = total momentum of the particles of the system.

Here, before the collision and in the LCS:

photon: $E_1 = h\nu \quad , \; p_1 = \frac{h\nu}{c}$

nucleus: $E_2 = Mc^2 \; , \quad p_2 = 0$

and, after the absorption of the photon, in the CMSC of the compound nucleus:

$$E = Mc^2 + E_e \; , \quad p = 0.$$

Equation (3) becomes then:

$$(h\nu + Mc^2)^2 - (h\nu)^2 = (Mc^2 + E_e)^2$$

or:

$$(h\nu)^2 + (Mc^2)^2 - (h\nu)^2 + 2Mc^2 h\nu = (Mc^2)^2 + 2E_e Mc^2 + E_e^2$$

$$\therefore \qquad E_e^2 + 2E_e\, Mc^2 - 2Mc^2\, h\nu = 0$$

Solving for E_e:

$$E_e = \frac{-2Mc^2 \pm \sqrt{(2Mc^2)^2 + 4(2Mc^2\, h\nu)}}{2}$$

and since only the + sign has physical meaning:

$$E_e = Mc^2 \left(\sqrt{1 + \frac{2h\nu}{Mc^2}} - 1\right) \qquad \text{as before.}$$

Discussion: the usual case will be when $h\nu \ll Mc^2$. Then, using the approximation:

$$\sqrt{1+x} = 1 + \frac{1}{2}x + \cdots\cdots \qquad \text{where:} \quad x = \frac{2h\nu}{Mc^2}$$

and neglecting higher powers of x, one has:

$$E_e = Mc^2 \left(1 + \frac{h\nu}{Mc^2} - 1\right) = h\nu$$

i.e., the excitation energy is just the energy of the incident photon.

****

98

A foil of Al^{27} is bombarded with a 200 microampere beam of 1 Mev protons. The foil is 0.05 mm thick, and the density of aluminum is 2.7 gm/cm^3. A proton-alpha reaction takes place with a cross section of 20 barns. Determine how long the foil should be bombarded in order to produce one microgram of the product element.

****

The nuclear reaction is written:

$$_{13}Al^{27} + {_1H^1} \rightarrow {_2He^4} + {_{12}Mg^{24}}$$

Let: n_t = total number of atoms in the foil = nAt

A = area of foil

t = thickness of foil = 5×10^{-3} cm

n = number of target atoms per cubic cm

N = number of protons incident on foil per unit time

σ = reaction cross section = 20 barns

The probability of interaction for one proton will be:

$$f = \frac{n_t\sigma}{A} = nt\sigma$$

where n is given by the equation:

$$n = \frac{N_o\rho}{M}$$

Here: N_o = Avogadro's number = 6.023×10^{23} atoms/gram atomic weight

M = atomic mass of target, gm/gram atomic weight

The reaction rate is obtained by multiplying the probability of interaction for one proton by the number of protons incident upon the foil per unit time:

$$R = Nf = N\frac{N_o}{M}\rho t\sigma$$

But R is also the number of atoms of the product isotope formed per unit time; thus, after a certain time T has elapsed, the number of atoms of Mg^{24} will be:

$$N_1 = RT$$

where, again, N_1 is given by the expression:

$$N_1 = \frac{N_oW}{M_p}$$

Here: W = mass of Mg formed (= 1 μgm)
M_p = atomic mass of Mg^{24}

Hence:

$$\frac{N_oW}{M_p} = RT = N\frac{N_o}{M}\rho t\sigma T$$

and solving for T:

$$\boxed{T = \frac{M}{M_p}\frac{W}{N\rho t\sigma}} \qquad (1)$$

The value of N is calculated as follows:

$$200\ \mu\text{amp} = 200 \times 10^{-6}\ \text{amp} = 2 \times 10^{-4}\ \frac{\text{coulomb}}{\text{sec}}$$

But the charge of the proton is:

$$e = 1.6 \times 10^{-19}\ \text{coulomb/proton}$$

Therefore:

$$N = \frac{2 \times 10^{-4}\ (\text{coul/sec})}{1.6 \times 10^{-19}\ (\text{coul/proton})} = 1.25 \times 10^{15}\ \frac{\text{protons}}{\text{sec}}$$

Replacing numerical values in (1), using CGS units, and taking the mass numbers instead of the atomic masses, one has:

$$T = \frac{27}{24}\ \frac{10^{-6}}{1.25 \times 10^{15} \times 2.7 \times 5 \times 10^{-3} \times 20 \times 10^{-24}}\ \text{sec}$$

or:

$$\boxed{T = 3333\ \text{sec} = 55.5\ \text{min}}$$

Answer

****

99 What is the excitation energy of the compound nucleus obtained by bombarding Cu^{63} with 6 Mev deuterons ?

Answer: $E_e = 19.16$ Mev

****

100 A 0.01 mm thick foil of ordinary aluminum of density 2.7 gram/cm^3 is bombarded with a 20 microampere beam of 1 Mev protons. Assume that only (p,n) reactions take place. Find the cross section in barns for this reaction if the rate of production of neutrons is 4×10^{11} per second.

Answer: $\sigma = 53$ barns.

****

7

ACCELERATORS.

MISCELLANEOUS.

101

A betatron is designed for an orbit radius of 40 cm. Electrons are injected at essentially zero energy when the magnetic flux density is zero and increasing, and extracted when B = 1990 gauss. Find the momentum, velocity, relativistic mass, and kinetic energy of the electrons at extraction.

****

The basic relation is obtained by equating the magnetic force and the centripetal force:

$$Bev = \frac{mv^2}{r}, \qquad \text{or:} \qquad \boxed{mv = Ber}$$

Thus, the momentum is:

$$p = Ber = 1990 \times \frac{4.8 \times 10^{-10}}{3 \times 10^{10}} \times 40$$

$$\therefore \qquad \boxed{p = 1.27 \times 10^{-15} \text{ dyne-sec}} \qquad \underline{\text{Ans}}.$$

Next, one calculates the total energy E:

$$E^2 = p^2c^2 + m_o^2c^4 \qquad (1)$$

Expressing pc in Mev:

$$pc = 1990 \times 4.8 \times 10^{-10} \times 40 \text{ (erg)} \times \frac{1 \text{ (Mev)}}{1.6 \times 10^{-6} \text{(erg)}}$$

$$\text{or:} \qquad pc = 23.85 \text{ Mev}$$

Using equation (1):

$$E = \sqrt{(23.85)^2 + (0.511)^2}$$

The numerical approximation can be improved, even if one uses a slide rule, by proceeding as follows:

$$E = 23.85\sqrt{1 + (\frac{0.511}{23.85})^2} = 23.85\sqrt{1 + (0.0214)^2}$$

$$= 23.85\sqrt{1 + 4.6 \times 10^{-4}} \cong 23.85\,(1 + \frac{1}{2} \times 4.6 \times 10^{-4})$$

$$= 23.85\,(1 + 0.00023)$$

Thus:

$$E = 23.85 \text{ Mev}$$

and the kinetic energy is: $K = 23.85 - 0.511$, or: $\boxed{K = 23.34 \text{ Mev}}$ <u>Ans</u>.

Now, from: $E = \gamma E_o$, one gets:

$$\gamma = \frac{E}{E_o} = \frac{23.83}{0.511} = 46.7$$

Then:

$$\beta = \left(1 - \frac{1}{\gamma^2}\right)^{1/2} \simeq 1 - \frac{1}{2\gamma^2} = 1 - \frac{1}{2 \times (46.7)^2} = 1 - 0.00023 = 0.99977$$

$$\text{or:} \qquad \boxed{v = 0.99977\,c} \qquad \underline{\text{Ans}}$$

i.e., the electrons are extreme relativistic.

Finally, the relativistic mass is:

$$m = \gamma m_o = 46.7 \times 9.1 \times 10^{-28} \text{ gm}$$

or: $$\boxed{m = 4.25 \times 10^{-26} \text{ gm}}$$ Ans.

****

102

Electrons are accelerated to an energy of 5 Mev in a linear accelerator, and then injected into a synchrotron of radius 13.25 m, from which they are extracted with an energy of 5 Gev. The energy gain per revolution is 1 Kev.

(a) Calculate the initial frequency of the RF source. Will it be necessary to change this frequency ?
(b) How many turns will the electrons make ?
(c) Calculate the time between injection and extraction of the electrons. What distance do the electrons travel within the synchrotron ?

****

(a) At injection: $\gamma - 1 = \frac{K}{E_o} = \frac{5}{0.511} = 9.78$ $\quad\therefore\quad \gamma = 10.78$

and: $\beta = 0.995$ (from tables)

At extraction: $\gamma - 1 = \frac{5 \times 10^3}{0.511} \simeq 10^4$ and: $\beta \simeq 1$

Now:

$$v = \omega r, \quad \text{but:} \left\{ \begin{array}{l} \omega = 2\pi\nu \\ v = \beta c \end{array} \right\} \quad \therefore \quad \beta c = 2\pi\nu r, \quad \text{or:} \quad \boxed{\nu = \frac{\beta c}{2\pi r}}$$

Numerically, then, the initial frequency is:

$$\nu = \frac{0.995 \times 3 \times 10^{10}}{2\ \pi \times 1325} = 3.58 \times 10^6 \text{ cycles/sec}$$

or: $\boxed{\nu = 3.58 \text{ MHz}}$ Ans.(a)

Since the final frequency would be 3.6 MHz, it will be not necessary to change the radiofrequency, which will remain practically constant.

(b) The total energy increase of the electrons is:

$$E_f - E_i = 5 \times 10^3 - 5 = 4995 \text{ Mev}$$

and since the energy gain per revolution is 1 Kev, the number of turns is:

$$n = \frac{4995}{1 \times 10^{-3}} \quad \text{or:} \quad \boxed{n = 4.995 \times 10^6 \text{ turns}}$$ Ans.(b)

(c) The time for one revolution is the period, $T = 1/\nu$; hence, the total time is:

$$t = T\,n = \frac{n}{\nu} = \frac{4.995 \times 10^6}{3.6 \times 10^6} \quad \text{or:} \quad \boxed{t = 1.39 \text{ sec}}$$ Ans(c)

Another way is to calculate the distance traveled:

$$d = 2\pi r\,n = 2\pi \times 1325 \times 4.995 \times 10^6 = 4.16 \times 10^{10} \text{ cm}$$

and divide by c: $t = \frac{4.16 \times 10^{10}}{3 \times 10^{10}} = 1.39$ sec, as before.

The total distance traveled is expressed also:

$$\boxed{d = 4.16 \times 10^5 \text{ Km}}$$ Ans(c)

****

103

The Stanford linear accelerator produces 60 pulses per second of about 10^{11} electrons, with a final energy of 1 Gev.
Calculate: (a) the average beam current; (b) the power output in watts; and (c), the average force on the target, assuming all the electrons are brought to rest.

****

(a) Beam current:

$$i = 60\left(\frac{\text{pulses}}{\text{sec}}\right) \times 10^{11}\left(\frac{\text{electrons}}{\text{pulse}}\right) \times 1.6 \times 10^{-19}\left(\frac{\text{coulombs}}{\text{electron}}\right)$$

$$= 6 \times 1.6 \times 10^{-7}\left(\frac{\text{coul}}{\text{sec}}\right) = 9.6 \times 10^{-7} \text{ amp}$$

or: $\boxed{i = 0.96\ \mu\text{amp}}$ Ans.(a)

(b) Power output:

$$W = 60\left(\frac{\text{pulses}}{\text{sec}}\right) \times 10^{11}\left(\frac{\text{elect.}}{\text{pulse}}\right) \times 10^{3}\left(\frac{\text{Mev}}{\text{elec.}}\right) \times 1.6 \times 10^{-13}\left(\frac{\text{joules}}{\text{Mev}}\right)$$

$$= 6 \times 1.6 \times 10^{2}\left(\frac{\text{joule}}{\text{sec}}\right) = 9.6 \times 10^{2} \text{ watts}$$

or: $\boxed{W = 960 \text{ watts}}$ Ans.(b)

(c) Average force on target:

Since the electrons are extreme relativistic:

$$K = E = pc, \quad \text{and then:} \quad p = \frac{E}{c} = \frac{K}{c}$$

By Newton's Second Law:

$$F = \frac{\Delta p}{\Delta t}$$

and if the electrons are brought to rest at the target: $\Delta p = p - 0 = p$

Then:

$$F = \frac{p}{\Delta t} = \frac{K}{c\Delta t}$$

But:

$$\frac{K}{\Delta t} = \frac{\text{energy}}{\text{unit time}} = \text{power} = W$$

so finally:

$$F = \frac{W}{c} = \frac{960(\text{watt})}{3 \times 10^{8}(\text{m/sec})} = 320 \times 10^{-8} \text{ newton}$$

or: $\boxed{F = 0.32 \times 10^{-5} \text{ newton} = 0.32 \text{ dyne}}$ Ans.(c)

****

104

Protons of 1 Mev energy enter a linear accelerator which has 99 drift tubes connected alternately to a 200 MHz oscillator. The final energy of the protons is 50 Mev.

(a) Calculate the initial and final velocity of the protons.
(b) What are the lengths of the second cylinder and the last cylinder?
(c) how many additional tubes would be needed to produce 60 Mev protons in this accelerator ?
(d) The 50 Mev accelerator produces 125 pulses/sec of 10^{10} protons each. What is the power output in watts ?

****

(a) 1 Mev protons are not relativistic and thus:

$$v = \sqrt{\frac{2K}{m}} = \left[\frac{2 \times 1 \times 1.6 \times 10^{-13}}{1.6725 \times 10^{-27}}\right]^{1/2} = 10^7 \sqrt{\frac{2 \times 1.6}{1.6725}} = 10^7 \sqrt{1.912}$$

or: $$v_i = 1.381 \times 10^7 \text{ m/sec}$$

which is also expressed: $$\boxed{v_i = 0.046c}$$ Ans.(a)

At the final energy, the protons are relativistic and:

$$\gamma = \frac{E}{E_o} = \frac{E_o + K}{E_o} = \frac{938.25 + 50}{938.25} = 1 + \frac{50}{938.25} = 1.053$$

$$\therefore \quad \beta = \sqrt{1 - \frac{1}{\gamma^2}} = \sqrt{1 - 0.900} = \sqrt{0.10} = 0.316$$

Hence:

$$\boxed{v_f = 0.316\ c}$$ Ans.(a)

or also: $$v_f = 0.9486 \times 10^8 \text{ m/sec}$$

(b) The accelerator has 98 gaps, so that the energy gain per gap is:

$$\Delta E = \frac{50-1}{98} = \frac{49}{98} = 0.5 \text{ Mev/gap}$$

After crossing the first gap, the protons are still non-relativistic and:

$$v_2 = \sqrt{\frac{2K}{m}} = \sqrt{\frac{2 \times (1+0.5) \times 1.6 \times 10^{-13}}{1.6725 \times 10^{-27}}} = 1.694 \times 10^7 \text{ m/sec}$$

The time within each drift tube is the same:

$$t = \frac{1}{2f} = \frac{1}{2 \times 2 \times 10^8} = 0.25 \times 10^{-8} \text{ sec}$$

The length of the second tube is therefore:

$$\ell_2 = v_2 t = 1.694 \times 10^9 \times 0.25 \times 10^{-8}$$

or: $$\boxed{\ell_2 = 4.24 \text{ cm}}$$ Ans.(b)

For the final tube, where the protons have already the final energy:

$$\ell_f = v_f t = 9.486 \times 10^9 \times 0.25 \times 10^{-8}$$

or: $\boxed{\ell_f = 23.71 \text{ cm}}$ Ans.(b)

(c) As already shown, the energy gain per gap is 0.5 Mev. Thus, to produce 60 Mev protons, 20 additional gaps will be necessary, or:

$\Delta N = 20$ additional tubes Ans.(c)

(d) The power output is:

$$W = 125 \left(\frac{\text{pulses}}{\text{sec}}\right) \times 10^{10} \left(\frac{\text{protons}}{\text{pulse}}\right) \times 50 \left(\frac{\text{Mev}}{\text{proton}}\right) \times 1.6 \times 10^{-13} \left(\frac{\text{joule}}{\text{Mev}}\right)$$

or: $$W = 125 \times 10^{10} \times 50 \times 1.6 \times 10^{-13} \text{ joule/sec}$$

$\therefore$ $\boxed{W = 10 \text{ watts}}$ Ans.(d)

*****

105

Positronium has a half-life of 10^{-10} seconds† and decays into two gamma-ray photons. Calculate the wavelength of these gamma-rays.

****

To conserve linear momentum the photons will be moving in opposite directions, and will carry each half of the energy resulting from the annihilation of the electron-positron pair, i.e., 0.511 Mev. Thus:

$$E_\gamma = h\nu = \frac{hc}{\lambda}$$

or:

$$\lambda = \frac{hc}{E_\gamma} = \frac{6.626 \times 10^{-27}\ (\text{erg-sec}) \times 3 \times 10^{10}\ (\text{m/sec}) \times 10^{8}(\text{Å/cm})}{0.511\ (\text{Mev}) \times 1.6 \times 10^{-6}\ (\text{erg/Mev})}$$

$$\therefore \quad \boxed{\lambda = 0.0243\ \text{Å}}$$

****

(†) For a discussion of the two possible types of positronium (ortho- and para-forms) see C.M.H. Smith, Nuclear Physics, Pergamon Press, 1966, p.695.

106

How close will an 8.776 Mev alpha particle get to a uranium nucleus in a head-on collision before being turned back ?

****

The alpha particle will turn back when all its kinetic energy has been converted into potential energy, i.e., when:

$$\frac{2Ze^2}{R} = E \qquad (1)$$

where: Z = atomic number of nucleus = 92
R = distance to center of nucleus

Thus:

$$R = \frac{2Ze^2}{E} = \frac{2 \times 92 \times (4.8 \times 10^{-10})^2}{8.776 \times 1.6 \times 10^{-6}} = 3.02 \times 10^{-12} \quad \text{cm}$$

or: $\boxed{R = 30.2 \text{ fermis}}$ Answer

The radius of the nucleus is given by:

$$r = 1.3 \quad A^{1/3} \text{ fermis}$$

and for U^{238}, one has:

$$r = 1.3 \times 6.197 = 8.05 \text{ fermis}$$

The minimum distance to the "surface" of the nucleus is then:

$$\Delta r = R - r = 30.2 - 8.05$$

or: $\boxed{\Delta r = 22.15 \text{ fermis}}$ Answer

****

107 The mean life of the first excited state of a certain nucleus is 1.3×10^{-14} seconds. Calculate the width of this energy level.

****

The mean life, τ, and the width, Γ, of the energy level are related by the equation:

$$\boxed{\Gamma\tau = \hbar = \frac{h}{2\pi}}$$

as follows from Heisenberg uncertainty principle.

Thus:

$$\Gamma = \frac{h}{2\pi\tau} = \frac{6.626 \times 10^{-27}}{2\pi \times 1.3 \times 10^{-14}} \text{ (erg)} \times \frac{1 \text{ (ev)}}{1.6 \times 10^{-12} \text{ (erg)}}$$

or: $\boxed{\Gamma = 0.0506 \text{ ev}}$ <u>Answer</u>

****

108 A neutral π meson of rest mass M moving with velocity βc in the LCS dissociates into two photons according to the reaction:

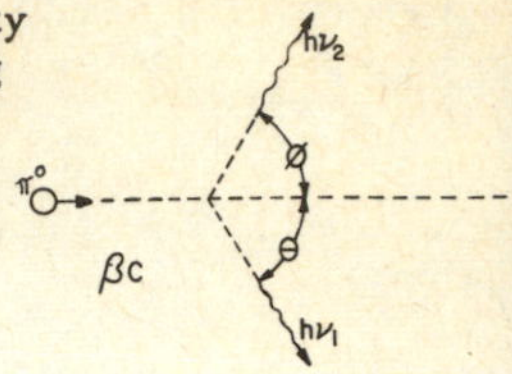

$$\pi \rightarrow h\nu_1 + h\nu_2$$

(a) Show that a photon emitted at an angle θ will have an energy:

$$h\nu_1 = \frac{Mc^2}{2\gamma(1 - \beta \cos \theta)}$$

where: $\gamma = 1/\sqrt{1-\beta^2}$

(b) Show that the maximum and minimum photon energies are given by:

$$h\nu_{max} = \frac{\gamma Mc^2}{2}(1+\beta) , \qquad h\nu_{min} = \frac{\gamma Mc^2}{2}(1-\beta)$$

****

(a) By conservation of energy:

$$\gamma Mc^2 = h\nu_1 + h\nu_2 \qquad (1)$$

By conservation of momentum:

$$\gamma\beta Mc = \frac{h\nu_1}{c}\cos\theta + \frac{h\nu_2}{c}\cos\phi$$

which is also written: $\gamma\beta Mc^2 = h\nu_1 \cos\theta + h\nu_2 \cos\phi$

and: $0 = h\nu_1 \sin\theta + h\nu_2 \sin\phi$

The angle ϕ is eliminated by squaring and adding the last two equations:

$$(\gamma\beta Mc^2)^2 + (h\nu_1)^2 - 2\gamma\beta Mc^2\, h\nu_1 \cos\theta = (h\nu_2)^2$$

and using equation (1):

$$(\gamma\beta Mc^2)^2 + (h\nu_1)^2 - 2\gamma\beta Mc^2\, h\nu_1 \cos\theta = (\gamma Mc^2)^2 + (h\nu_1)^2 - 2\gamma Mc^2\, h\nu_1$$

Dividing by γMc^2:

$$\gamma Mc^2\beta^2 - 2\beta\, h\nu_1 \cos\theta = \gamma Mc^2 - 2h\nu_1$$

and solving for $h\nu_1$:

$$h\nu_1 = \frac{\gamma Mc^2(1-\beta^2)}{2(1 - \beta\cos\theta)}$$

and since: $\gamma(1-\beta^2) = 1/\gamma$, finally:

$$\boxed{h\nu_1 = \frac{Mc^2}{2\gamma(1-\beta\cos\theta)}} \qquad (2) \qquad \text{Q.E.D.}$$

(b) From equation (1), the energy of one photon is maximum when the energy of the other is minimum. Now, from (2), the maximum energy is obtained when the denominator is minimum, i.e., when $\cos\theta = 1$. Then:

$$h\nu_{max} = \frac{Mc^2}{2\gamma(1-\beta)} = \frac{Mc^2(1+\beta)}{2\gamma(1-\beta^2)} = \frac{\gamma Mc^2}{2}(1+\beta)$$

Similarly, the photon energy is minimum when the denominator is maximum, ie., when $\cos\theta = -1$. Then:

$$h\nu_{min} = \frac{Mc^2}{2\gamma(1+\beta)} = \frac{Mc^2(1-\beta)}{2\gamma(1-\beta^2)} = \frac{\gamma Mc^2}{2}(1-\beta) \qquad \text{Q.E.D.}$$

****

109

A hypothetical nucleus of mass $M = 4.6978$ amu at rest spontaneously splits into two parts of equal mass m. Each part has kinetic energy equal to 1/100 of its rest energy.

(a) What fraction of the mass of the original nucleus was converted into kinetic energy ?

(b) Calculate the kinetic energy of each product nucleus in Mev.

****

(a) Conservation of energy is expressed:

$$Mc^2 = mc^2 + \frac{mc^2}{100} + mc^2 + \frac{mc^2}{100}$$

or:

$$M = m + \frac{m}{100} + m + \frac{m}{100} = \frac{202}{100}\, m$$

Thus:

$$m = \frac{100}{202} M \qquad (1)$$

Now:

$$Q = (\text{initial masses}) - (\text{final masses}) = Mc^2 - 2mc^2$$

and using (1):

$$Q = Mc^2 - \frac{200}{202} Mc^2 = \frac{2Mc^2}{202}$$

The fraction converted into kinetic energy is therefore:

$$f = \frac{Q}{Mc^2} = \frac{2}{202} \qquad \text{or:} \qquad \boxed{f = 0.0099 \simeq 1\%} \qquad \underline{\text{Ans.(a)}}$$

(b) The kinetic energy of each product nucleus is:

$$K = \frac{1}{100} mc^2 = \frac{1}{100} \frac{100}{202} Mc^2 = \frac{4.6978 \text{ (amu)} \times 931.5 \text{ (Mev/amu)}}{202}$$

$$\text{or:} \qquad \boxed{K = 21.6 \text{ Mev}} \qquad \underline{\text{Ans. (b)}}$$

****

110 Prove that in a head-on collision of a heavy particle with an electron the maximum velocity acquired by the latter is twice the velocity of the incident particle.

****

Let: M = mass of heavy particle, moving with initial velocity v_o
m = mass of electron, at rest before the collision

By conservation of momentum:

$$Mv_o = Mv_1 + mv_2 \qquad (1)$$

By conservation of energy:

$$\frac{1}{2} Mv_o^2 = \frac{1}{2} Mv_1^2 + \frac{1}{2} mv_2^2 \qquad (2)$$

From equation (1):

$$M(v_o - v_1) = mv_2$$

or:

$$m = M \frac{v_o - v_1}{v_2}$$

By substitution into (2):

$$Mv_o^2 = Mv_1^2 + M \frac{v_o - v_1}{v_2} v_2^2$$

or:

$$v_o^2 = v_1^2 + (v_o - v_1)v_2$$

Solving for v_2, the velocity of the electron after the collision:

$$v_2 = \frac{v_o^2 - v_1^2}{v_o - v_1} = v_o + v_1$$

Now, in the optimum case when $M >> m$, the velocity of the incident particle will not change appreciably, and:

$$v_1 \simeq v_o$$

Thus, at the most:

$$\boxed{v_2 = 2v_o}$$

Q.E.D.

****

111

Show that the maximum velocity that can be imparted to a proton at rest by a non-relativistic alpha particle is 1.6 times the velocity of the incident alpha particle.

****

Let v be the velocity of the incident alpha particle, and assume as a sufficient approximation that $m_\alpha = 4m$, where m is the proton mass. The optimum case is when the proton moves after the collision in the direction of the incident alpha. Conservation of momentum is written then:

$$4mv = 4mv_1 + mv_2$$

or:

$$v_1 = v - \frac{1}{4} v_2 \qquad (1)$$

By conservation of energy:

$$\frac{1}{2}(4m)v^2 = \frac{1}{2}(4m)v_1^2 + \frac{1}{2} mv_2^2$$

or:

$$2mv^2 = 2mv_1^2 + \frac{1}{2} mv_2^2$$

Substituting v_1 here, as given by (1), one has:

$$2v^2 = 2(v - \frac{1}{4} v_2)^2 + \frac{1}{2} v_2^2$$

$$\therefore \quad 2v^2 = 2v^2 - vv_2 + \frac{1}{8} v_2^2 + \frac{1}{2} v_2^2$$

or:

$$vv_2 = (\frac{1}{8} + \frac{1}{2})v_2^2 = \frac{5}{8} v_2^2$$

and finally:

$$\boxed{v_2 = \frac{8}{5} v = 1.6 \, v} \qquad \text{Q.E.D.}$$

****

112

A mass m with velocity v_o (v_o<<c) collides head-on with a mass M initially at rest. Show that if the collision is perfectly elastic, the energy E transferred to M is given by:

$$E = \frac{4mME_o}{(M+m)^2}, \qquad \text{where:} \quad E_o = \frac{1}{2} mv_o^2$$

****

Conservation of energy is written:

$$\frac{1}{2} mv_o^2 = \frac{1}{2} MV^2 + \frac{1}{2} mv^2 \qquad (1)$$

m → M before (v_o)

m → M after (v, V)

while conservation of momentum is expressed:

$$mv_o = MV + mv \qquad (2)$$

or:

$$mv = mv_o - MV$$

Replacing in (1), after multiplying by m:

$$\frac{1}{2} m^2v_o^2 = \frac{1}{2} mMV^2 + \frac{1}{2} (mv_o - MV)^2$$

$$= \frac{1}{2} mMV^2 + \frac{1}{2} m^2v_o^2 + \frac{1}{2} M^2V^2 - mMv_oV$$

or:

$$M^2V^2\left(\frac{m}{2M} + \frac{1}{2}\right) = mv_o MV$$

$$MV = \frac{2mMv_o}{M + m}$$

The final energy of M is now:

$$E = \frac{1}{2} MV^2 = \frac{(MV)^2}{2M} = \frac{4m^2M^2v_o^2}{2M(m+M)^2}$$

or: $$\boxed{E = \frac{4mME_o}{(M+m)^2}}$$ Q.E.D.

****

113

If ionizing radiation loses 32 ev in producing a single ion-pair, and if the density of air at STP is 1.293×10^{-3} gm/cm^3, show that 1 roentgen is equivalent to 2.58×10^{-4} coulombs/Kg air, and also to 82.5 ergs per gram of air.

****

By definition:

$$1 \text{ r} = 1 \frac{\text{esu}}{\text{cm}^3 \text{ air}}$$

but:

$$1 \text{ coulomb} = 3 \times 10^9 \text{ esu}$$
$$1 \text{ cm}^3 \text{air} = 1.293 \times 10^{-3} \text{ grams of air}$$

$$\therefore \quad 1 \text{ r} = \frac{1}{3 \times 10^9} \frac{1}{1.293 \times 10^{-3}} \frac{\text{coulombs}}{\text{gm of air}}$$

or:

$$\boxed{1 \text{ r} = 2.58 \times 10^{-4} \text{ coul/Kg air}} \qquad \text{Q.E.D.} \qquad (1)$$

To express this now in terms of energy per gram of air, it is necessary to use the experimentally determined value:

$$w = 32 \frac{\text{ev}}{\text{ion-pair}}$$

which is also expressed:

$$w = 32 \times \frac{1.6 \times 10^{-19}(\text{joule/ev})}{1.6 \times 10^{-19}(\text{coul/ion-pair})} = 32 \frac{\text{joule}}{\text{coul}}$$

Hence, 32 joules are deposited per coulomb of charge of either sign produced, and in (1):

$$1 \text{ r} = 2.58 \times 10^{-4}\left(\frac{\text{coul}}{\text{Kg air}}\right) \times 32 \left(\frac{\text{joule}}{\text{coul}}\right) \times 10^7\left(\frac{\text{erg}}{\text{joule}}\right) \times \frac{1}{10^3} \left(\frac{\text{Kg}}{\text{gm}}\right)$$

or:

$$\boxed{1 \text{ r} = 82.5 \frac{\text{ergs}}{\text{gm air}}} \qquad \text{Q.E.D.}$$

****

APPENDIX

FUNDAMENTAL PHYSICAL CONSTANTS

Name	Symbol	Value*	
Atomic mass unit	amu	1.6605×10^{-24}	gram
		931.4812	Mev
Avogadro's number	N	6.0221×10^{23}	$\frac{\text{molecules}}{\text{mole}}$
Bohr radius	$a_o = \hbar^2/m_e e^2$	5.2917×10^{-9}	cm
Bohr magneton	$\mu_B = e\hbar/2m_e c$	9.2740×10^{-21}	erg/gauss
Boltzmann's constant	$k = R_o/N$	1.3806×10^{-16}	erg/°K
		8.617×10^{-5}	ev/°K
Classical electron radius	$r_o = e^2/m_e c^2$	2.8179×10^{-13}	cm
Compton wavelength of the electron	$\lambda_C = h/m_e c$	2.4263×10^{-10}	cm
Electron charge	e	1.6021×10^{-19}	coulomb
		4.8032×10^{-10}	esu
Electron rest mass	m_e	9.1095×10^{-28}	gram
		0.511004	Mev
Fine-structure constant	$\alpha = e^2/\hbar c$	7.2973×10^{-3}	
	α^{-1}	137.0360	
Gas constant	R_o	8.3143×10^{7}	erg/mole °K
Gravitational constant	G	6.6732×10^{-8}	cm^3/sec^2 gram
Neutron rest mass	M_n	1.6749×10^{-24}	gram
		939.5527	Mev
Nuclear magneton	$\mu_n = e\hbar/2M_p c$	5.0509×10^{-24}	erg/gauss
Plank's constant	h	6.6261×10^{-27}	erg-sec
	$\hbar = h/2\pi$	1.0545×10^{-27}	erg-sec
Proton rest mass	M_p	1.6726×10^{-24}	gram
		938.2592	Mev
Rydberg constant	$R_\infty = m_e e^4/4\pi\hbar^3 c$	1.0973×10^{5}	cm^{-1}
Speed of light	c	2.9979×10^{10}	cm/sec
Standard volume of ideal gas	V_o	22.4136×10^{3}	cm^3/mole

(*) From Taylor et al., Revs. Mod. Phys., 41,3,375 (July 1969)

INDEX *

(*) Numbers refer to problems, not to pages.

******